FORMULA HANDBOOK FOR ENVIRONMENTAL ENGINEERS AND SCIENTISTS

ENVIRONMENTAL SCIENCE AND TECHNOLOGY

A Wiley-Interscience Series of Texts and Monographs

Edited by JERALD L. SCHNOOR, *University of Iowa*
ALEXANDER ZEHNDER, *Swiss Federal Institute for Water Resources and Water Pollution Control*

A complete list of the titles in this series appears at the end of this volume.

FORMULA HANDBOOK FOR ENVIRONMENTAL ENGINEERS AND SCIENTISTS

GABRIEL BITTON
Department of Environmental Engineering Sciences
University of Florida
Gainesville, Florida

A Wiley-Interscience Publication
JOHN WILEY & SONS, INC.
New York / Chichester / Weinheim / Brisbane / Singapore / Toronto

WISSER MEMORIAL LIBRARY
REF
TD153
.B58
1998
Copy 1

REFERENCE BOOK
NOT TO BE TAKEN
FROM THE LIBRARY

This text is printed on acid-free paper. ♾

Copyright © 1998 by John Wiley & Sons, Inc.

Published simultaneously in Canada.

No part of this publication may be reproduced, stored in a retrieval system or transmitted in any form or by any means, electronic, mechanical, photocopying, recording, scanning or otherwise, except as permitted under Sections 107 or 108 of the 1976 United States Copyright Act, without either the prior written permission of the Publisher, or authorization through payment of the appropriate per-copy fee to the Copyright Clearance Center, 222 Rosewood Drive, Danvers, MA 01923, (508) 750-8400, fax (508) 750-4744. Requests to the Publisher for permission should be addressed to the Permissions Department, John Wiley & Sons, Inc., 605 Third Avenue, New York, NY 10158-0012, (212) 850-6011, fax (212) 850-6008, E-Mail: PERMREQ@WILEY.COM.

Library of Congress Cataloging in Publication Data:

Bitton, Gabriel.
Formula handbook for environmental engineers and scientists/
Gabriel Bitton.
p. cm.—(Environmental science and technology)
Includes index.
ISBN 0-471-13905-X (cloth: alk. paper)
1. Environmental engineering—Mathematics—Handbooks, manuals,
etc. 2. Environmental sciences—Mathematics—Handbooks, manuals,
etc. 3. Engineering mathematics—Formulae—Handbook, manuals, etc.
I. Title. II. Series.
TD153.B58 1997
628—dc21 97-12769

Printed in the United States of America

10 9 8 7 6 5 4 3 2 1

To Sally and Billy Glick

SERIES PREFACE

Environmental Science and Technology

We are in the third decade of the Wiley-Interscience Series of texts and monographs in Environmental Science and Technology. It has a distinguished record of publishing outstanding reference texts on topics in the environmental sciences and engineering technology. Classic books have been published here, graduate students have benefited from the textbooks in this series, and the series has also provided for monographs on new developnments in various environmental areas.

As new editors of this Series, we wish to continue the tradition of excellence and to emphasize the interdisciplinary nature of the field of environmental science. We publish texts and monographs in environmental science and technology as it is broadly defined from basic science (biology, chemistry, physics, toxicology) of the environment (air, water, soil) to engineering technology (water and wastewater treatment, air pollution control, solid, soil, and hazardous wastes). The series is dedicated to a scientific description of environmental processes, the prevention of environmental problems, and preservation and remediation technology.

There is a new clarion for the environment. No longer are our pollution problems only local. Rather, the scale has grown to the global level. There is no such place as "upwind" any longer; we are all "downwind" from somebody else in the global environment. We must take care to preserve our resources as never before and to learn how to internalize the cost to prevent environmental degradation into the product that we make. A new "industrial ecology" is emerging that will lessen the impact our way of life has on our surroundings.

In the next 50 years, our population will come close to doubling, and if the developing countries are to improve their standard of living as is needed, we will require a gross world product several times what we currently have. This will create new pressures on the environment, both locally and globally. But there are new opportunities also. The world's people are recognizing the need for sustainable development and leaving a legacy of resources for future generations at least equal to what we had. The goal of this series is to help understand the environment, its functioning, and how problems can be overcome; the Series will also provide new insights and new sustainable technologies that will allow us to preserve and hand down an intact environment to future generations.

Jerald L. Schnoor
Alexander J. B. Zehnder

PREFACE

In the field of environmental engineering sciences, we are often faced with the problem of finding formulas and equations that cover a wide range of disciplines. Unfortunately, sometimes our failing memories do not allow the rapid retrieval of the formulas and their sources. We must then bear the burden of consulting countless books and journal articles to retrieve the appropriate piece of information on Monod's equation, the ATP content of a cell, the water–octanol partition coefficient, the Eadie–Scatchard plot, or the Lotka–Volterra equation. A glance at the formulas in this book will convince the reader about the interdisciplinary nature of the field of environmental engineering sciences.

This handbook should be a very useful addition to the library of scientists or engineers working in the environmental field. The book will save precious time in the tedious task of searching for formulas and equations in the ocean of literature. The book will serve both students and researchers in the disparate disciplines involved in environmental engineering sciences. In a modest way, this book should complement the traditional *Handbook of Physics and Chemistry* that has been part of our libraries for so many years.

In am grateful to colleagues who have encouraged me to undertake this task, and to the many publishing firms that have allowed the use of figures and tables displayed in this book. I would like to thank Professor Ben Koopman, who provided moral support and allowed me the use of his private book collection on wastewater engineering. I thank Eric Kuo for drawing most of the illustrations for this book. I am grateful to Nancy, Julie, Natalie, and all my family around the world for their unconditional love and support in this endeavor.

Gabriel Bitton

Gainesville, Florida
July, 1997

LIST OF TABLES

Table A1: α Factors for some Aeration Devices
Table A2: Effect of Temperature on K_b for Ammonia
Table A3: Autotrophic Indexes for Selected Samples
Table B1: Conversion Factors for Thymidine Incorporation in Bacterial Production
Table B2: Bioconcentrations factors for Some Hydrophobic Compounds
Table B3: Biomass of Microorganisms in Coastal Waters
Table B4: Biovolume of *P. putida* Cells According to Culture Conditions
Table B5: Estimates of Mean Cell Volumes in Aquatic Habitats
Table B6: Biovolumes of Some Ciliates from Florida Lakes
Table B7: Calculated Mean Cell Volumes of Representative Species of Freshwater and Marine Organisms
Table B8: Conversion Factors of Bacterial Biovolumes to Nitrogen Biomass
Table B9: Brownian displacements and Diffusion Coefficients Calculated for Uncharged Spheres in Water at 20°C
Table B10: Bulk Densities of Soils
Table C1: Chlorophyll *a* in Lakes and Reservoirs
Table C2: C/N Ratios of Bacterioplankton
Table D1: Hydraulic Conductivity of Some Soil and Aquifer Materials
Table D2: Decay Rates of Microorganisms in Some Environments
Table D3: Values of Lethality Coefficient λ and Residual O_3 to Destroy 99% of Microorganisms in 10 min at 10–15°C
Table D4: Values of λ for 99% Destruction of Four Groups of Organisms in 1 min at 50°C by Ozone and Three Chlorine Compounds
Table D5: Doubling Times of Some Bacterial Cultures

Table D6: Doubling Times of Bacteria in Marine Environments
Table F1: Typical FC/FS Ratios
Table F2: Food to Microorganisms Ratios in Some Activated Sludge Systems
Table F3: Free Energies of Hydrolysis of Some Phosphorylated Compounds
Table F4: Freundlich Isotherm: Values of K_f and $1/n$ for Some Priority Organic Pollutants at Neutral pH
Table G1: Growth yields Y of Some Microorganisms
Table H1: Henry's Law Constants for Some Organic Compounds
Table M1: Range of MLSS in Some Activated Sludge Processes
Table M2: Growth Kinetics at 15°C of Some Bacteria Isolated from Drinking Water
Table M3: μ_{max} and Doubling Times of Some Fungi Grown on Glucose
Table M4: Maximum Growth Rates of Some Protists and Algaue in Culture
Table M5: Most Probable Number (MPN) Index and 95% Confidence Limits When Five 20-mL portions Are Used
Table M6: Most Probable Number (MPN) Index and 95% Confidence Limits When Ten 10-mL Portions Are Used
Table M7: Most Probable number (MPN) Index and 95% Confidence Limits When Five Tubes Are Used per Dilution (10, 1.0, 0.1 mL of Sample)
Table N1: E_0 for Some Chemical Reactions
Table N2: Average N:P Mass Ratios in Potential Nutrient Sources of Freshwater Lakes
Table N3: Average N:P Mass Ratios in Aquatic Organisms
Table O1: Critical Oxygen Concentrations, C_{crit}, of Some Microorganisms
Table O2: Specific Oxygen Requirements of Pure Cultures of Microorganisms
Table P1: S_w, K_{ow}, and K_{oc} of Some Organic Chemicals
Table P2: Phenol Coefficients of Some Chlorinated Compounds (Using *S. typhi*)
Table P3: Porosity of Soils and Soil Constituents
Table P4: Factors for Conversion of Total Alkalinity to Milligrams of Carbon per Liter
Table P5: Primary productivity Rates for Phytoplankton in Fresh Waters
Table P6: Gross Primary Productivity of Estuaries Compared to Other Aquatic and Terrestrial Areas
Table P7: Net Primary Productivity in Limnic Ecosystems
Table P8: Relationship Between Percent Response, Normal Equivalents Deviations (NED), and Probit Units
Table P9: Productivity/Biomass (P/B) Ratios of Freshwater Organisms
Table Q1: Q_{10} for Selected Enzymes
Table S1: Typical Sludge Quantities Generated by Domestic Wastewater Treatment Plants

Table S2: Specific Gravity of Selected Solids
Table S3: Relationship Between the Equilibrium Constant K'_{eq} and Standard Free Energy Change at 25°C
Table S4: Standard Oxido-reduction Potentials E'_0 for Some Components of the Electron Transport Chain
Table S5: Surface Tension of Some Liquids
Table T1: Temperature Coefficients for Biological Processes
Table W1: Water Activity a_w Limits for Growth of Selected Microorganisms

LIST OF FIGURES

Figure A1: Effect of pH and Temperature on the Distribution of NH_3 and NH_4 in Water
Figure A2: Arrhenius Plot for Determining Activation Energy
Figure B1: Empirical Relationship Between Bacterial and Algal Production Across a Trophic Gradient
Figure B2: Bacterial Production in Lake Nesjovatn, Norway
Figure B3: Lambert's Law: Transmission of Light as a Function of the Thickness *z* of the Absorbing Medium
Figure B4: BET Isotherms: (A) Typical; (B) Linear B.E.T.
Figure B5: Biochemical Oxygen Demand (BOD): (a) Oxidation of Organic Matter; (b) Influence of Nitrification
Figure B6: BOD Curve: Consumption of Oxygen During Organic Matter Degradation
Figure B7: Biofilm
Figure C1: First-Order Kinetics for Chick's Law
Figure C2: Disinfection: Deviation from First-Order Kinetics
Figure C3: Chemostat for Continuous Culture of Microorganisms
Figure D1: Vertical Cross Section of Groundwater Flow with Linear, Parallel Streamlines
Figure D2: Determination of the *D* Value Concerning the Survival of a Spore former at 121°C
Figure D3: Effect of *n* value on *Ct* at Various Disinfectant Concentrations
Figure D4: Dixon Plots for (A) Competitive and (B) Noncompetitive Enzyme Inhibition
Figure E1: Eadie–Scatchard Plot

Figure E2: Electrical Double Layer of a Negatively Charged Colloidal Particle
Figure E3: Lineweaver–Burk Plot for Competitive Inhibition of Enzymes
Figure E4: Lineweaver–Burk Plot for Noncompetitive Inhibition of Enzymes
Figure E5: Lineweaver–Burk Plot for Uncompressible Inhibition of Enzymes
Figure E6: Growth Curve for Microorganisms
Figure F1: Changes in Free Energy During Exergonic and Endergonic Reactions
Figure F2: Plot of Freundlich Isotherm
Figure G1: Theoretical Population Growth in the Absence (J Curve) and Presence (S Curve) of Limiting Factors
Figure H1: Haldane Equation: Comparison of the Monod and Haldane Plots
Figure H2: Hanes Plot
Figure H3: Lineweaver–Burk Plot for Measuring Heterotrophic Potential
Figure H4: Hofstee Plot
Figure L1: Langmuir Isotherms: (a) Typical; (b) Linear
Figure L2: Compensation Light Intensity in Aquatic Environments
Figure L3: Light Intensity Versus Depth of the Water
Figure L4: Typical Lineweaver–Burk Plot
Figure M1: Median Lethal Concentration (LC_{50})
Figure M2: Michaelis–Menten Plot
Figure M3: Determination of Numerical Aperture of a Microscope Lens
Figure M4: Mathematical Relations Commonly Encountered in Modeling
Figure M5: Relationship Between Microbial Specific Growth Rate and Substrate Concentration According to Monod's Equation
Figure N1: Specific Decay Rates as a Function of Temperature and Mean Solids Retention Time
Figure P1: Determination of LC_{50} in Toxicity Testing via the Probit Units Method
Figure R1: Relationship Between the Specific Growth Rate μ and the Cell RNA/DNA Ratio
Figure S1: Typical Dissolved Oxygen Sag Curve
Figure T1: Influence of Trickling Filter Depth and Hydraulic Loading Intensity on Substrate Removal Rate

INTRODUCTION

The field of environmental engineering sciences has blossomed within the past three decades, and is multidisciplinary by nature. It draws upon a myriad of disciplines in the biological, chemical, and physical sciences, as well as political science, sociology, and economics. Engineers and scientists working in the environmental field must by necessity have a basic understanding of biological, physical, and chemical processes occurring in the environment and must often consult the literature in disparate disciplines ranging from microbiology, biochemistry, meteorology, wastewater engineering, hydrodynamics, aerosol mechanics, ecology, limnology, chemistry, physics, soil science, hydrology, geology, and toxicology, to statistics or thermodynamics. Years of experience have taught me that working in this eclectic field necessitates knowledge of or the ability for rapid retrieval of key equations and formulas in the myriad of disciplines that serve as a basis of environmental engineering and sciences.

In this book I have covered formulas that deal primarily with biological/biochemical processes in natural and engineered systems. Numerous journal articles, review articles, books, and handbooks were consulted to select the most useful and popular formulas.

For each formula, I have attempted to supply the following type of information:

- *Definition or introduction*: The subject is introduced briefly or a given term or process is defined.
- *Formula*: The formula is given and the units are indicated.
- *Some numerical values reported in the literature*: Whenever possible, numerical values found in the literature are given. This information is of

utmost importance to some researchers (e.g., modelers), who will be able to retrieve these values or range of values rapidly and use them in their models.
- *References*: Pertinent references are supplied for each formula. When possible, the original reference is included.
- When appropriate, the text is illustrated with pertinent tables and figures.

The formulas are presented in alphabetical order and some of them are cross-referenced. The formulas were not organized according to the discipline covered because some of them belong in various categories, which makes their retrieval more difficult.

The most popular linear models are given for several formulas (e.g., Linewaver–Burk plot, Hanes plot, Eadie–Scatchard plot). Some linearization models have given biased estimates of parameters, however; a better approach to estimating these parameters is nonlinear least squares analysis (Berthouex and Brown, 1994). The reader is reminded that software for nonlinear parameter estimation is widely available. Parameter estimation requires in-depth knowledge of the models. Hence the user should also consult the original reference before applying parameter estimation techniques.

In the appendixes, useful conversion tables are supplied. These tables go hand in hand with formulas and equations and constitute useful pieces of information.

References

Berthouex, P. M., and L. C. Brown, 1994. *Statistics for Environmental Engineers*, Lewis Publishers, Boca Raton, FL.

FORMULA HANDBOOK FOR ENVIRONMENTAL ENGINEERS AND SCIENTISTS

A

ABSORPTION COEFFICIENT

See LIGHT: Vertical Extinction Coefficient.

ACTIVATED SLUDGE: Floc Load

Definition

In activated sludge plants, floc load is the organic loading over a specified time period at the point where influent wastewater or primary effluent mixes with return activated sludge.

Formula

$$\underset{\text{(mg COD/g MLSS)}}{\text{Floc load}} = \frac{\text{mass of available COD in unit time at initial mixing}}{\text{mass of MLSS in unit time at initial mixing}}$$

where

COD = chemical oxygen demand [BOD (biochemical oxygen demand) can be used instead of COD]

MLSS = mixed liquor suspended solids

References

Eikelbloom, D. H. 1982. Biosorption and prevention of bulking sludge by means of a high floc loading. In: *Bulking of Activated Sludge: Preventive and Remedial Measures*, B. Chambers and E. J. Tomlinson, Eds., Ellis Horwood, Chichester, West Sussex, England.

Institute of Water Population Control. 1987. *Unit Processes: Activated Sludge*, IWPC, Maidstone, Kent, England.

ACTIVATED SLUDGE: Oxygen Consumption Rate

Definition

The oxygen consumption rate is the amount of oxygen consumed by 1 g of activated sludge per hour. It is a useful parameter under both laboratory and field conditions.

Formula

$$\underset{(\text{mg/g}\cdot\text{h})}{\text{Oxygen consumption rate}} = \frac{\text{oxygen consumption rate (mg/L/min)}}{\text{volatile suspended solids}} \times \frac{60\ \text{min}}{\text{h}}$$

Reference

American Public Health Association. 1989. *Standard Methods for the Examination of Water and Wastewater*, 17th ed., APHA, Washington, DC.

ADDITIVE INDEX

See TOXICITY OF CHEMICAL MIXTURES: Additive Index.

ADENOSINE TRIPHOSPHATE (ATP): Relationship with Microbial Biomass

Introduction

Adenosine triphosphate (ATP) in cells provides a good index of living biomass in environmental samples. ATP concentration is determined by the firefly luminescence assay.

Formulas and Numerical Values

ATP Content of Cells

ATP Content/Cell

- Laboratory-grown *Bacillus* sp.: 3×10^{-15} g ATP/cell (Moriarty, 1977).
- Laboratory-grown *Pseudomonas*: 0.5×10^{-15} g ATP/cell (Moriarty, 1977).

- For marine bacteria, Holm-Hansen and Booth (1966) estimated an ATP content of 1.5×10^{-9} g ATP/cell.
- Subsurface bacteria (Cape Cod, Massachusetts): $13.70 \pm 6.04 \times 10^{-18}$ g ATP/cell (Metge et al., 1993).

ATP/g Dry Weight of Cells

- Knowles (1977) reported that the ATP content of aerobically or anaerobically growing heterotrophic or photosynthetic bacteria was in the range 2–10 μmol ATP/g dry weight.
- 0.05–0.12 mg/g dw of cells in biofilms (Gikas and Livingston, 1993).
- 0.65–1.7 mg/g dw in activated sludge (Chiu et al., 1973).
- For marine bacteria, Holm-Hansen and Booth (1966) estimated an ATP content of 1–2 mg/g dw of cells, while for marine algae, the content varied between 0.03 and 1.6 mg/g dw of cells.

ATP/DNA Ratio This ratio was proposed as an index of metabolic activity of microorganisms (Jeffrey and Paul, 1986). Some examples are:

- Subsurface bacteria
  Cape Cod: 0.001–0.003 (Metge et al., 1993)
  Oklahoma: <0.001–0.018 (Webster et al., 1985)
- *Vibrio proteolytica* (laboratory grown): 0.18–0.26 (Jeffrey and Paul, 1986)

Conversion of ATP to Dry Weight of Cells For most microorganisms, ATP levels generally range from 1 to 4 μg/mg dry weight (dw) (this range can be between 0.03 and 12 μg/mg dw (Weber, 1973). A value of 2.4 μg ATP/mg dw organic matter is assumed (American Public Health Association, 1989; Weber, 1973).

Conversion of ATP to Organic Carbon In aquatic environments, live microbial biomass can generally be derived from ATP value according to the following equation (Holm–Hansen and Booth, 1966):

$$\mathrm{ATP} = \frac{\mathrm{C}}{250}$$

where ATP is in mg/L and C is organic carbon (mg/L). However, Herbert (1990) stressed that this conversion factor was obtained using Tris buffer extraction and that a change in the extraction procedure may lead to an altered ratio.

In soils, Oades and Jenkinson (1979) proposed the following relationship, which might not hold for all soils:

$$\text{ATP} = \frac{\text{C}}{120}$$

Others reported that the relationship between biomass carbon and ATP in soils was 10.6 μmol ATP/g biomass C (Jenkinson et al., 1979; Tate and Jenkinson, 1982). Brookes and Ocio (1989) reported 9.2 μmol ATP/g biomass C.

References

American Public Health Association. 1989. *Standard Methods for the Examination of Water and Wastewater*, 17th ed., APHA, Washington, DC.

Atlas, R. M., and R. Bartha, 1981. *Microbial Ecology: Fundamentals and Applications*, Addison-Wesley, Reading, MA.

Brookes, P. C., and J. A. Ocio, 1989. The use of ATP measurements in soil microbial biomass studies, pp. 129–136. In: *ATP Luminescence: Rapid Methods in Microbiology*, P. E. Stanley, B. J. McCarthy, and R. Smither, Eds., Blackwell Scientific, Oxford.

Chiu, S. Y., I. C. Kao, L. E. Erickson, and L. T. Fan. 1973. ATP pools in activated sludge. J. Water Pollut. Control Fed. 45: 1746–1756.

Gikas, P., and A. G. Livingston. 1993. Use of ATP to characterize biomass viability in freely suspended and immobilized cell bioreactors. Biotechnol. Bioeng. 42: 1337–1351.

Herbert, R. A. (1990). Methods for enumerating microorganisms and determining biomass in natural environments. Methods Microbiol. 22: 1–39 (R. Grigorova and J. R. Norris, Eds.), Academic Press, London.

Holm-Hansen, O., and C. R. Booth. 1966. The measurement of adenosine triphosphate in the ocean and its ecological significance. Limnol. Oceanogr. 11: 510–519.

Jeffey, W. H., and J. H. Paul. 1986. Activity of an attached and free-living *Vibrio* sp. as measured by thymidine incorporation, *p*-iodonitrotetrazolium reduction, and ATP/DNA ratios. Appl. Environ. Microbiol. 51: 150–156.

Jenkinson, D. S., S. A. Davidson, and D. S. Powlson. 1979. Adenosine triphosphate and microbial biomass in soil. Soil Biol. Biochem. 11: 521–527.

Knowles, C. J. 1977. Microbial metabolic regulation by adenine nucleotide pools, pp 241–283. In: *Microbial Energetics*, B. A. Haddock and W. A. Hamilton, Eds., Cambridge University Press, Cambridge.

Metge, T. W., M. H. Brooks, R. L. Smith, and R. W. Harvey, 1993. Effect of treated-sewage contamination upon bacterial energy charge, adenine nucleotides, and DNA content in a sandy aquifer on Cape Cod. Appl. Environ. Microbiol. 59: 2304–2310.

Moriarty, D. J. W. 1977. Improved method using muramic acid to estimate biomass of bacteria in sediments. Oecologia 26: 317–323.

Oades, J. M., and D. S. Jenkinson. 1979. Adenosine triphosphate content of the soil microbiol biomass. Soil Biol. Biochem. 11: 201–204.

Tate, K. R., and D. S. Jenkinson. 1982. Adenosine triphosphate in soil: an improved method. Soil. Biol. Biochem. 14: 331–336.

Weber, C. I. 1973. Recent developments in the measurement of the response of plankton and periphyton to changes in their environment, pp. 119–138. In: *Bioassay Techniques and Environmental Chemistry*, G. Glass, Ed., Ann Arbor Science, Ann Arbor, MI.

Webster, J. J., G. J. Hampton, J. T. Wilson, W. C. Ghiorse, and F. R. Leach. 1985. Determination of microbial cell numbers in subsurface samples. Ground Water 23: 17–25.

ADENYLATE ENERGY CHARGE (AEC)

Definition

Adenylate energy charge is an index used as a measure of the metabolic state of microbial populations.

Formula

$$\mathrm{AEC} = \frac{\mathrm{ATP} + \frac{1}{2}\mathrm{ADP}}{\mathrm{ATP} + \mathrm{ADP} + \mathrm{AMP}}$$

where

ATP = adenosine triphosphate
ADP = adenosine diphosphate
AMP = adenosine monophosphate

Numerical Values

According to Chapman et al. (1971),

- AEC = 0.8–0.9 indicates active microbial growth.
- AEC = 0.5–0.6 indicates stationary phase and senescence.

Aquifers In aquifer samples from Cape Cod, EC = 0.58 ± 0.01 (indicate metabolic stress for aquifer microorganisms; some of the microorganisms were senescent or in the stationary phase) (Metge et al., 1993).

Sediments AEC = 0.22–0.32 for streambed sediments (Kaplan and Bott, 1985). AEC = 0.19–0.49 for marine sediments (Christensen and Devol, 1980).

For marine sediments in upwelling zone off the coast of northwestern Africa, AEC = 0.53–0.7 in the upper sediment layer (Tan and Ruger, 1989).

Lakes For the bloom of the diatom *Melosira* in Lake Pavin (France), AEC = 0.61–0.85 (Amblard and Bourdier, 1990).

Soils For organic soil with 4.29% C (Brookes and Jenkinson, 1989; Brookes et al., 1983):

- Fresh soil: AEC = 0.85–0.87
- Air-dried: AEC = 0.45
- Air-dried and remoistened soil: AEC = 0.76
- Soil incubated with ryegrass for 50 days: AEC = 0.90

AEC of agricultural soils (Munich, Germany) is 0.56–0.71 (Zelles et al., 1994).

References

Amblard, C., and G. Bourdier. 1990. The spring bloom of the diatom *Melosira italica* subsp. *subarctica* in Lake Pavin: biochemical, energetic and metabolic aspects during sedimentation. J. Plankton Res. 12: 645–660.

Atkinson, D. E. 1977. *Cellular Energy Metabolism and Its Regulation*, Academic Press, San Diego, CA.

Atlas, R. M., and R. Bartha. 1981. *Microbial Ecology: Fundamentals and Applications.* Addison-Wesley, Reading, MA.

Brookes, P. C., and D. J. Jenkinson. 1989. ATP and adenylate energy charge levels in the soil microbial biomass, pp. 119–127. In: *ATP Luminescence: Rapid Methods in Microbiology*, P. E. Stanley, B. J. McCarthy, and R. Smither, Eds., Blackwell Scientific, Oxford.

Brookes, P. C., K. R. Tate, and D. J. Jenkinson. 1983. The adenylate energy charge of the soil microbiol biomass. Soil Biol. Biochem. *15: 9–16.*

Chapman, A. G., L. Fall, and D. E. Atkinson. 1971. J. Bacteriol. 108: 1072–1086.

Christensen, J. P., and A. H. Devol. 1980. Adenosone triphosphate and energy charge in marine sediments. Mar. Biol. 56: 175–182.

Kaplan, L. A., and T. L. Bott. 1985. Adenylate energy charge in streambed sediments. Freshwater Biol. 15: 133–138.

Metge, T. W., M. H. Brooks, R. L. Smith, and R. W. Harvey. 1993. Effect of treated-sewage contamination upon bacterial energy charge, adenine nucleotides, and DNA content in a sandy aquifer on Cape Cod. Appl. Environ. Microbiol. 59: 2304–2310.

Tan, T. L., and H.-J. Ruger. 1989. Benthic studies of the northwest African upwelling region: bacteria standing stock and ETS activity, ATP-biomass and adenylate energy charge. Mar. Ecol. Prog. Ser. 51: 167–176.

Wiebe, W. J., and K. Bancroft. 1975. Use of the adenylate energy charge ratio to measure growth state of natural microbial communities. Proc. Nat Acad. Sci. USA 72: 2112–2115.

Zelles, L., Q. Y. Bai, R. X. Ma, R. Rackwitz, K. Winter, and F. Beese. 1994. Microbial biomass, metabolic activity and nutritional status determined from fatty acid patterns and poly-hydroxybutyrate in agriculturally-managed soils. Soil Biol. Biochem. 26: 439–446.

ADENYLATE POOL (A_T)

Introduction

Adenylate pool is used as a measure of activity of microorganisms. It is not dependent on the metabolic state of the microorganism.

Formula

$$A_T = \text{ATP} + \text{ADP} + \text{AMP}$$

where

ATP = adenosine triphosphate
ADP = adenosine diphosphate
AMP = adenosine monophosphate

Numerical Values

Aquifer bacteria (Cape Cod, Massachusetts): $A_T = 6.02 \pm 1.04 \times 10^{-20}$ mol/cell (average for seven sites) (Metge et al., 1993).

References

Atlas, R. M., and R. Bartha, 1981. *Microbial Ecology: Fundamentals and Applications*, Addison-Wesley, Reading, MA.

Metge, T. W., M. H. Brooks, R. L. Smith, and R. W. Harvey. 1993. Effect of treated-sewage contamination upon bacterial energy charge, adenine nucleotides, and DNA content in a sandy aquifer on Cape Cod. Appl. Environ. Microbiol. 59: 2304–2310.

ADSORPTION ISOTHERM

Introduction

An adsorption isotherm expresses the adsorption as a function of the concentration of the adsorbate in bulk solution at constant temperature. The amount of adsorbed material per unit weight of adsorbent generally increases with the adsorbate concentration (Weber, 1972). Various theoretical and empirical models have been proposed to describe the various types of adsorption isotherms (*see* BET ISOTHERM; FREUNDLICH ISOTHERMS; LANGMUIR ISOTHERM).

References

Weber, W. J. Jr. 1972. *Physicochemical Processes for Water Quality Control.* Wiley-Interscience, N.Y., 640 pp.

AEROSOL, BIOLOGICAL: Predictive Model

Introduction

Dispersion models borrowed from the field of aerosol mechanics and requiring the integration of biological and meteorological data have been developed to predict the downwind concentrations of aerosolized microorganisms from known sources (e.g., activated sludge units, spray irrigation sites) and to assess the possible health significance of microbial aerosols. These models assume a Gaussian (normal) distribution of pollutants at any specific downwind location (Mohr, 1991; Reed et al., 1988; U.S. EPA, 1982).

Formulas

The downwind concentration C_d of aerosolized microorganisms is given by the following equation:

$$C_d = C_n D_d e^{xa} + B$$

where

C_n = microbial concentration at the source (number/m^3)

D_d = atmospheric dispersion factor (s/m^3)

x = microbial decay rate (s^{-1})

a = aerosol age (downwind distance/wind speed) (s)

B = background concentration (number/m^3)

The microbial concentration at the source C_n is given by the following formula.

$$C_n = WFEI$$

where

W = microbial concentration in wastewater (number/L)

F = flow rate of wastewater (L/s)

E = aerosolization efficiency (s/m^3) ($E = 0.3\%$ for wastewater; $E = 0.04$ for sludge applied via spray gun)

I = survival factor of aerosolized microorganisms (I varies with the pathogen under consideration)

References

Mohr, A. J. 1991. Development of models to explain the survival of viruses and bacteria in aerosols, pp. 160–190. In: *Modeling the Environmental Fate of Microorganisms*, C. J. Hurst, Ed., American Society for Microbiology, Washington, DC.

Reed, S. C., E. J. Middlebrooks, and R. W. Crites. 1988. *Natural Systems for Waste Management and Treatment*, McGraw-Hill, New York.

U.S. EPA. 1982. *Estimating Microorganisms Densities in Aerosols from Spray Irrigation of Wastewater*, EPA-600/9-82-003, Center for Environmental Research Information, Cincinnati, OH.

ALPHA-FACTOR

Definition

The alpha factor is the ratio of the overall volumetric mass transfer coefficient for oxygen ($K_L a$) in wastewater to the $K_L a$ measured in clean water.

Formula

$$\alpha = \frac{K_L a \text{ wastewater}}{K_L a \text{ clean water}}$$

where $K_L a$ is expressed in units of 1/time.

Numerical Values

- $\alpha = 0.3$ in domestic wastewater at the beginning of the aeration period in an activated sludge unit.

TABLE A1: α Factors for Some Aeration Devices

Aeration Device	α Factor	Wastewater
Fine bubble diffuser	0.4–0.6	Tap water containing detergent
Coarse bubble diffuser, sparger	0.7–0.8	Domestic wastewater
Coarse bubble diffuser, sparger	0.55	Activated sludge contact tank
Surface aerators	0.6–1.2	Tap water with detergent and small amounts of activated sludge

Source: Adopted from Eckenfelder (1989).

- $\alpha = 0.8$ after 4 h of aeration.
- $\alpha = 0.79$ after 3 h of aeration of kraft mill wastes.

α varies with the aeration device (Table A1).

References

Eckenfelder, W. W. Jr. 1989. *Industrial Water Pollution Control*, 2nd ed., McGraw-Hill, New York.

Institute of Water Pollution Control. 1987. *Unit Processes: Activated Sludge*, IWPC, Maidstone, Kent, England.

AMMONIA, FREE

Introduction

Ammonia exists in solution in two forms, the free or gaseous form (NH_3) and the ionized form NH_4^+). The proportion of the free form is important because this species is toxic to the biota.

Formula and Numerical Values

$$NH_3\% = \frac{100}{1 + K_b[H^+]/K_w}$$

TABLE A2: Effect of Temperature on K_b for Ammonia

Temperature (°C)	$K_b \times 10^5$
0	1.374
10	1.570
20	1.710
30	1.820
40	1.862
50	1.892

Source: Schroeder (1977).

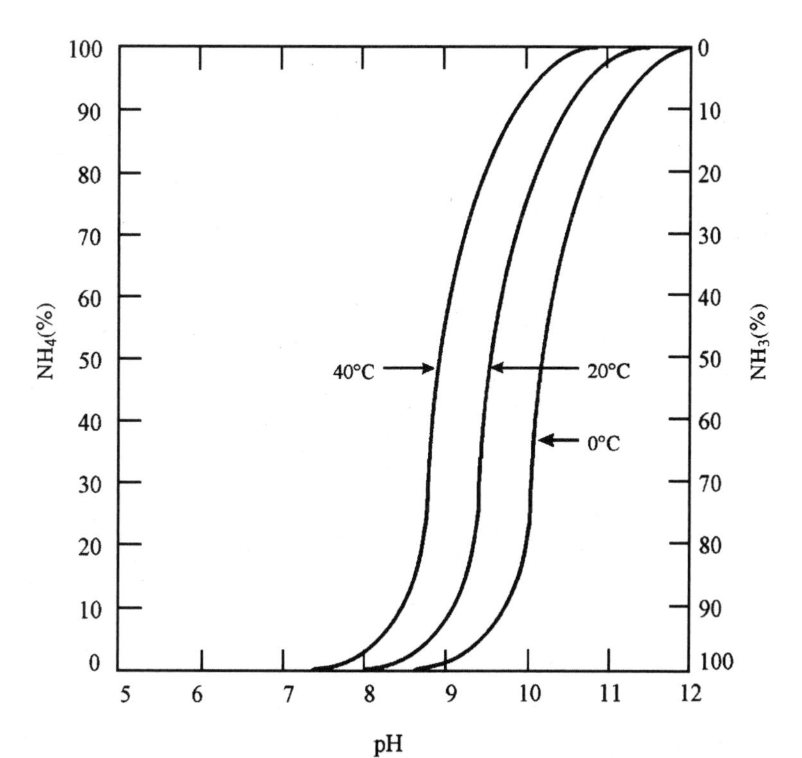

FIGURE A1: Effect of pH and Temperature on the Distribution of NH_3 and NH_4 in Water (*Source:* U.S. EPA, 1975.)

where

K_w = ion product of water = 1×10^{-14} at 25°C
= $[H^+][OH^-]$

K_b = equilibrium coefficient

$$= \frac{[NH_4{}^+][OH^-]}{[NH_3]}$$

K_b is a function of temperature (Table A2).

$[NH_3]$ depends on pH, with $[NH_3]$ increasing as pH increases. The relationship between pH and % NH_3 is shown in Figure A1. The figure also shows the effect of temperature on % NH_3.

References

Benefield, L. D., J. F. Judkins, Jr., and B. L. Weand. 1982. *Process Chemistry for Water and Wastewater Treatment*, Prentice Hall, Upper Saddle River, NJ.

Schroeder, E. D. 1977. *Water and Wastewater Treatment*, McGraw-Hill, New York.

U.S. EPA. 1975. *Process Design Manual for Nitrogen Control*, U.S. Government Printing Office, Washington, DC.

APPLICATION FACTOR (AF)

Definition/Introduction

In environmental toxicology, the application factor (AF) is the ratio of the maximum acceptable toxicant concentration (MATC) to the median lethal concentration (LC_{50}). MATC is obtained via chronic toxicity tests, whereas LC_{50} is obtained via acute toxicity tests. AF is relatively constant for a given chemical.

Formula

$$AF = \frac{MATC}{LC_{50}}$$

where

MATC = maximum acceptable toxicant concentration (mg or μg/L)

LC_{50} = median lethal concentration (mg or μg/L)

Sometimes, the *acute/chronic ratio* (ACR) is used and is the reciprocal of the application factor (Macek, 1985).

Numerical Values

In effluent monitoring, the application factor AF is used to extrapolate from LC_{50} concentrations to no-effect concentration, and ranges between 0.1 and 0.01, depending on the persistence and bioaccumulation potential (higher AF is allowed when the waste does not persist or bioaccumulate) (Mount, 1977; NAS/NAE, 1972; Tebo, 1986). The reported range for pulp and paper mill effluents was 0.05–0.1 (Walden, 1976).

References

Macek, K. J. 1985. Effluent evaluation, pp. 636–649. In: *Fundamentals of Aquatic Toxicology*, G. M. Rand and S. R. Petrocelli, Eds., Hemisphere, Washington, DC.

Mount, D. I. 1977. *An Assessment of Application Factors in Aquatic Toxicology*, EPA-600/3-77-085, U.S. Environmental Protection Agency, Washington, DC.

Mount, D. I., and C. E. Stephan. 1967. A method for establishing acceptable limits for fish: malathion and the butoxyethanol ester of 2,4-D. Trans. Am. Fish Soc. 96: 185–193.

NAS/NAE (National Academy of Sciences and National Academy of Engineering). 1972. *Water Quality Criteria*, 1972, EPA-R3-73-033, U.S. Environmental Protection Agency, Washington, DC.

Rand, G. M., and S. R. Petrocelli, Eds. 1985. *Fundamentals of Aquatic Toxicology*, Hemisphere, Washington, DC.

Tebo, L. B., Jr. 1986. Effluent monitoring: historical perspective, pp. 13–31. In: *Environmental Hazard Assessment of Effluents*, H. L. Bergman, R. A. Maki, and A. W. Maki, Eds., Pergamon Press, Elmsford, NY.

Walden, C. C. 1976. The toxicity of pulp and paper mill effluents and corresponding measurement procedures. Water Res. 10: 639–664.

ARRHENIUS EQUATION

Introduction

In 1889, a Swedish chemist, Arrhenius, proposed an equation to describe the effect of temperature on enzymatic reactions and chemical reactions in general. This equation is also used to describe the effect of temperature on the death rate of microorganisms.

Formula and Numerical Values

$$K = Ae^{-(E/RT)}$$

The linear form of this equation is

$$\ln K = -\frac{E}{R}\frac{1}{T} + A$$

where

K = specific rate constant for the enzymatic or chemical reaction (day^{-1})
E = activation energy (cal/mol)
R = gas constant (1.987 cal/K · mol)
T = absolute temperature (K)
A = empirical constant

The Arrhenius equation is used for the determination of the rate constant K at several temperatures. A plot of ln K versus $1/T$, generally called the *Arrhenius plot*, is linear with a slope of $-E/R$ (Figure A2). Experimental values of E range from a few kcal/mol to more than 25 kcal (Stumm and Morgan, 1981).

The effect of temperature on microbial growth or thermal decay is also described by the Arrhenius equation (Trilli, 1986).

$$K = Ae^{-(E/RT)}$$

where

K = thermal decay rate or growth rate (h^{-1})
A = empirical constant
E = activation energy (cal/mol)
r = gas constant (1.987 cal/K · mol)
T = absolute temperature (K)

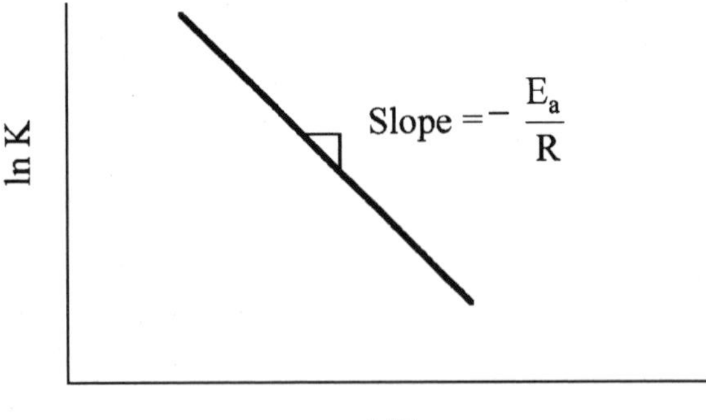

FIGURE A2: Arrhenius PLot for Determining Activation Energy (*Source:* Benefield et al., 1982.)

For example, for the thermal inactivation of spores of *Bacillus stearothermophilus*, the following values were obtained for A and E: $A = 1 \times 10^{36.2}\,s^{-1}$; $E = 67.7\,kcal/mol$ (Deindoerfer and Humphrey, 1961).

References

Arrhenius, S. 1889. Uber die Reaktionsgeschwindigkeit bei der Inversion von Rohrzucker durch Säuren. Z. Phys. Chem. 4: 226–248.

Benefield, L. D., J. F. Judkins, Jr., and B. L. Weand. 1982. *Process Chemistry for Water and Wastewater Treatment*, Prentice-Hall, Upper Saddle River, NJ.

Deindoerfer, F. H., and A. E. Humphrey. 1961. Scale-up of heat sterilization operations. Appl. Microbiol. 9: 134–139.

Gaudy, A. F., Jr., and E. T. Gaudy. 1988. *Elements of Bioenvironmental Engineering*, Engineering Press, San Jose, CA.

Grady, C. P. L., and H. C. Lim. 1980. *Biological Wastewater Treatment: Theory and Applications*, Marcel Dekker, New York.

Stumm, W., and J. J. Morgan. 1981. *Aquatic Chemistry*, Wiley-Interscience, New York.

Trilli, A. 1986. Scale-up fermentations, pp. 277–307. In: *Manual of Industrial Microbiology and Biotechnology*, A. L. Demain and N. A. Solomon, Eds., American Society for Microbiology, Washington, DC.

AUTOHEATED THERMOPHILIC AEROBIC DIGESTION: Rise in Temperature

Introduction

Autoheated thermophilic aerobic digestion is a process that generates heat as a result of free energy released by the aerobic degradation of organic matter by sludge microorganisms. Much of the energy resulting from the oxidation of organic compounds is released as heat. The rise in temperature is a function of the level of organic matter.

Formula

The rise in temperature is given by the following equation (Forster and Senior, 1987; Jewell and Kabrick, 1980):

$$\Delta T = 2.4\Delta(\mathrm{COD})$$

where

ΔT = increase in temperature (°C)

$\Delta(\mathrm{COD})$ = change in chemical oxygen demand (mg/L)

References

Forster, C. F., and E. Senior. 1987. Solid waste, pp. 176–233. In: *Environmental Biotechnology*, C. F. Forster and D. A. J. Wase, Eds., Ellis Horwood, Chichester, West Sussex, England.

Jewell, W. J., and R. M. Kabrick. 1980. Autoheated aerobic thermophilic digestion with aeration. J. Water Pollut. Control Fed. 52: 512–523.

AUTOTROPHIC INDEX

Definition/Introduction

The autotrophic index (AI), the ratio of total biomass to chlorophyll *a*, is a water quality indicator. This ratio increases when the water is enriched with organic matter, a situation that leads to an increase in the numbers of heterotrophic microorganisms (e.g., bacteria, fungi, protozoa).

Formula

$$AI = \frac{\text{total biomass (ash-free weight of organic matter)(mg/m}^3\text{)}}{\text{chlorophyll } a \text{ (mg/m}^3\text{)}}$$

Numerical Values

Some AI values are reported in Table A3. The normal AI values is 50–200.

TABLE A3: Autotropic Index for Selected Samples

Sample	Autotrophic Index
Algal culture	40–96
Marine phytoplankton	76–200
Pond Water	44–221
Marine seston	40–146
Lake seston	457

Source: adapted from Weber (1973)

References

American Public Health Association. 1989. *Standard Methods for the Examination of Water and Wastewater*, 17th ed., APHA, Washington, DC.

Weber, C. I. 1973. Recent developments in the measurement of the response of plankton and periphyton to changes in their environment, pp. 119–138. In: *Bioassay Techniques and Environmental Chemistry*, G. Glass, Ed., Ann Arbor Science, Ann Arbor, MI.

B

BACTERIAL PRODUCTION

Introduction

Bacteria feed on organic exudates produced by algae, the primary producers. Bacterial secondary production is related to phytoplankton primary production (Hobbie and Cole, 1984; Cole et al., 1988; Cole and Caraco, 1993). Bacterial production becomes less important when the primary production increases (Figure B1).

Formula

$$BP = CF \times (0.347 \times NPP^{0.8})$$

where

- BP = bacterial production (mg C/L · day)
- NPP = algal net primary production (mg C/L · day)
- CF = correction factor = 1.56

Another general formula for bacterial production in aquatic environments is (Riemann, 1983)

$$\text{bacterial production} = B\mu$$

where

- μ = specific growth rate of bacteria (h^{-1})
- B = bacterial biomass (mg C/m^3)
  = total bacterial biovolume × (1.21×10^{-13} g C)

Numerical Values

- Bacterial production in Mirror Lake, New Hampshire, in the epilimnion (summer): BP = 5–8 mg C/L · day (mean = 6.8 ± 1.3) (Cole and Caraco, 1993).
- In the surface (0–2 cm) sediment of a hypereutrophic lake in Sweden, the bacterial production at 20°C was 12–28 μg C/g dry weight per hour. Bacterial production varied with the season, ranging from 0.25–0.45 to 12–22 μg C/g dry weight per hour in August (Bell and Ahlgren, 1987).

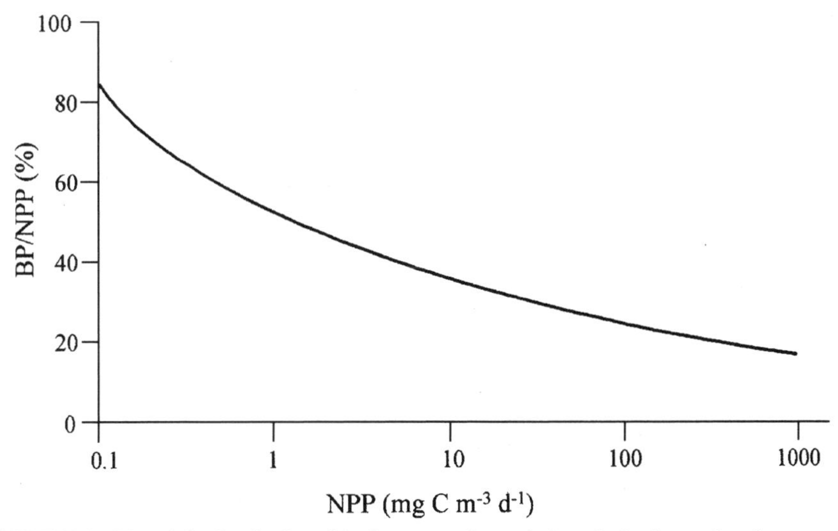

FIGURE B1: Empirical relationship between baterial and algal production across a Trophic gradient.

- Average bacterial production in Lake Nesjovatn, Norway (eutrophic lake) = 170 mg C/m^2 · day (range of 0–450 mg C/m^2 · day). This represents 25% of primary production on a seasonal basis. Bacterial production from April to October 1981 in Lake Nesjovatn is displayed in Figure B2.
- Bacterial production in sediments from various environments ranges from 10.5 to 500 mg C/m^2 · day (results normalized to 1 cm depth for comparison) (Moriarty, 1986).
- Saanich Inlet, British Columbia, Canada (June and August 1978): 6.6–71 mg C/L · day (Fuhrman and Azam, 1980).
- McMurdo Sound, Antarctica (December–January): 0.0004–2.9 mg C/L · day (Fuhrman and Azam, 1980).
- Scripps Pier, La Jolla, California (March, May, and August): 0.7–13.0 mg C/L · day (Fuhrman and Azam, 1980).

Comparison of Primary Production to Bacterial Production A summary of the literature revealed that in marine environments, bacterial production varied from 0.2 to 5600 ng C/L · h and represented from 5 to 45% of primary productivity. It has been suggested that bacterial growth uses at least 25% of primary production (Moriarty, 1986). In three freshwater lakes, bacterial production varied from 18 to 45% of primary production (data summarized by Riemann, 1983).

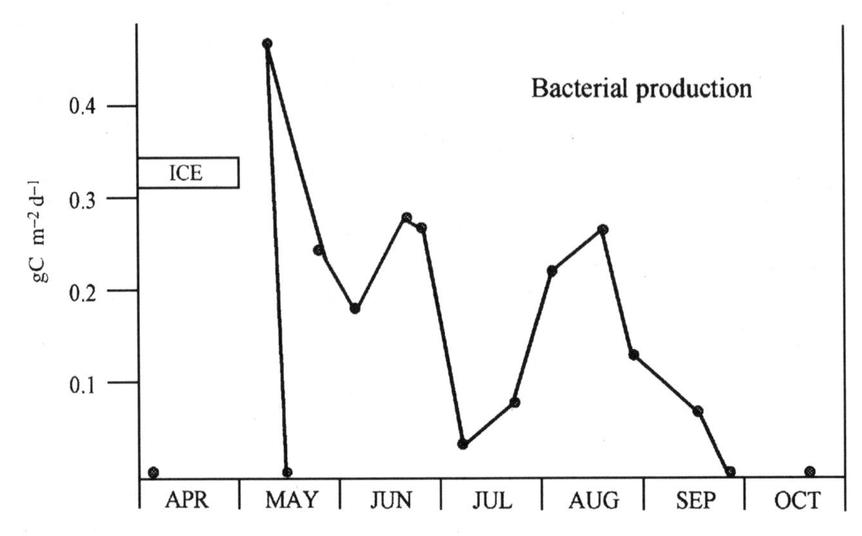

FIGURE B2: Bacterial production in Lake Nesjovatn, Norway.

Conversion Factors Between Bacterial Production and Thymidine Incorporation into Bacterial DNA (Staley and Konopka, 1985) [^{3}H]thymidine incorporation into bacterial DNA is now accepted as a tool for determining bacterial production in environmental samples (e.g., seawater, fresh water, sediments). However, the calculations are based on several assumptions (Riemann et al., 1987):

TABLE B1: Conversion Factors for Thymidine Incorporation in Bacterial Production

Method[a]	Value (10^{18} cells/mol thymidine)	Reference
A	1.7–2.4	Fuhrman and Azam (1982)
	1.3	Moriarty and Pollard (1981)
	1.1	Findlay et al. (1984)
B	1.4	Fuhrman and Azam (1982)
	2.0	Bell et al. (1983)
	1.1	Riemann et al. (1987)
	4.7–18.3	Scavia et al. (1986)
C	3.0–68	Kirchman et al. (1982)

[a] A, theoretical calculation; B, correlation with measured increases in cell numbers; C, correlation with rate of change of thymidine-incorporation rate.

- Thymidine constitutes 25% of the DNA bases.
- DNA is a constant fraction of cell carbon.
- Thymidine is incorporated exclusively into cell DNA.

The conversion factors of thymidine incorporation to bacterial numbers are summarized in Table B1.

Note: According to Riemann et al. (1987), the high conversion factors reported by Kirchman et al. (1982) are due to the assumptions that growth was exponential and that all bacterial cells were active (a lag phase and 60 to 80% inactive bacteria were observed by other investigators). From a survey of the literature, Moriarty (1986) proposed a conversion factor of 2×10^{18}.

References

Bell, R. T., and I. Ahlgren. 1987. Thymidine incorporation and microbial respiration in the surface sediment of a hypereutrophic lake. Limnol. Oceanogr. 32: 476–482.

Bell, R. T., G. M. Ahlgren, and I. Ahlgren. 1983. Estimating bacterioplankton production by measuring [^{3}H]thymidine incorporation in a eutrophic Swedish lake. Appl. Environ. Microbiol. 45: 1709–1721.

Cole, J. J., and N. F. Caraco. 1993. The pelagic microbial food web of oligotrophic lakes, pp. 101–111. In: *Aquatic Microbiology: An Ecological Perspective*, T. E. Ford, Ed., Blackwell Scientific, Oxford.

Cole, J. J., S. Findlay, and M. L. Pace. 1988. Bacterial production in fresh and saltwater ecosystems: a cross-system overview. Mar. Ecol. Prog. Ser. 43: 1–10.

Findlay, S. E. G., J. L. Meyer, and R. T. Edwards. 1984. Measuring bacterial production via rate of incorporation of [^{3}H]thymidine into DNA. J. Microbiol. Methods 2: 57–72.

Fuhrman, J. A., and F. Azam. 1980. Bacterioplankton secondary production estimates for coastal waters of British Columbia, Antarctica and California. Appl. Environ. Microbiol. 39: 1085–1095.

Fuhrman, J. A., and F. Azam. 1982. Thymidine incorporation as a measure of heterotrophic bacterioplankton production in marine surface waters: evaluation and field results. Mar. Biol. 66: 109–120.

Hobbie, J. E., and J. J. Cole. 1984. Response of a detrital foodweb to eutrophication. Bull. Mar. Sci. 35: 357–363.

Kirchman, D., H. Ducklow, and R. Mitchell. 1982. Estimates of bacterial growth from changes in uptake rates and biomass. Appl. Environ. Microbiol. 44: 1296–1307.

Moriarty, D. J. W. 1986. Measurement of bacterial growth rates in aquatic systems from rates of nucleic acid synthesis. Adv. Microb. Ecol. 9: 245–292.

Moriarty, D. J. W., and P. C. Pollard. 1982. DNA synthesis as a measure of bacterial production in seagrass sediments. Mar. Ecol. Prog. Ser. 5: 151–156.

Riemann, B. (1983). Biomass and production of phyto- and bacterioplankton in eutrophic Lake Tystrup, Denmark. Freshwater Biol. 13: 389–398.

Riemann, B., P. K. Bjornsen, S. Newell, and R. Fallon. 1987. Calculation of cell production of coastal marine bacteria based on measured incorporation of [^{3}H]thymidine. Limnol. Oceanogr. 32: 471–476.

Scavia, D., G. A. Laird, and G. L. Fahnenstiel. 1986. Production of planktonic bacteria in Lake Michigan. Limnol. Oceanogr. 31: 612–626.

Staley, J. T., and A. Konopka. 1985. Measurement of *in situ* activities of nonphotosynthetic microorganisms in aquatic and terrestrial habitats. Annu. Rev. Microbiol. 39: 321–346.

Vadstein, O., B. O. Harkjerr, A. Jensen, Y. Olsen, and R. Reinertsen. 1989. Cycling of organic carbon in the photic zone of a eutrophic lake with special reference to the heterotrophic bacteria. Limnol. Oceanogr. 34: 840–855.

BEER–LAMBERT LAW

See also LIGHT: Vertical Extinction Coefficient.

Introduction

The Beer–Lambert law describes the absorption of a monochromatic light by a solute, using a spectrophotometer. The intensity of a monochromatic light entering an absorbing medium decreases exponentially with the thickness as well as the concentration of the absorbing medium.

Formulas

Lambert's Law Formula The intensity of the transmitted light I is a function of the thickness z of the absorbing medium and is given by (Figure B3).

$$I = I_0 e^{-\alpha z}$$

$$\ln \frac{I_0}{I} = \alpha z$$

where

I_0 = intensity of incident light

I = intensity of transmitted light

α = absorption coefficient, which depends on the extinction coefficient of the medium

z = thickness of the absorbing medium (cm)

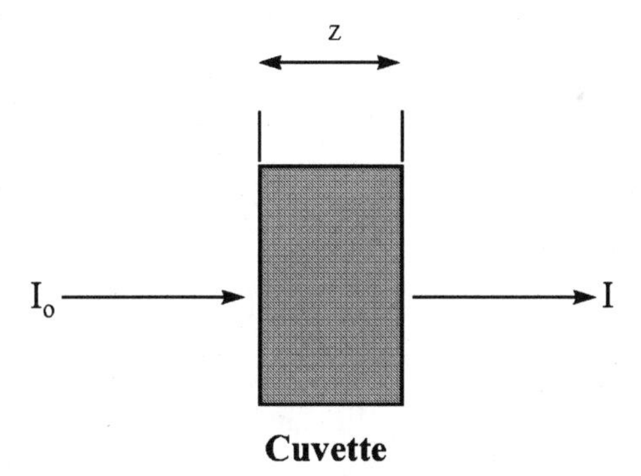

Cuvette

FIGURE B3: Lambert's Law: transmission of light as a function of the thickness *z* of the absorbing medium.

Using logarithms to the base 10, the absorption coefficient α is converted into the extinction coefficient K:

$$\alpha = 2.303K$$

Thus the absorbance A is given by

$$A = \log_{10}\frac{I_0}{I} = Kz$$

Beer's Law Formula Beer's law deals with light absorption as a function of solution concentration. Absorbance A is given by:

$$A = \log\frac{I_0}{I} = k''C$$

where

k'' = index of absorption

C = concentration of absorbing medium (mol/l)

Beer–Lambert Law Formula (combination of Beer and Lambert laws)

$$A = \log\frac{I_0}{I} = kCz$$

where k is the molar extinction coefficient ($cm^2/\mu mol$). The Beer–Lambert relationship is valid only for dilute solutions.

References

Clark, J. M., and R. L. Switzer. 1977. *Experimental Biochemistry*, W.H. Freeman, San Francisco.

Lind, O. W. 1979. *Handbook of Common Methods in Limnology*, 2nd ed., C.V. Mosby, St. Louis, MO.

Sawyer, C. N., P. L. McCarty, and G. F. Parkin. 1994. *Chemistry for Environmental Engineering*, McGraw-Hill, New York.

Sukatsch, D. A., and A. Dziengel. 1987. *Biotechnology: A Handbook of Practical Formulae*, Longman, Harlow, Essex, England.

Wetzel, R. G., and G. E. Likens. 1991. *Limnological Analyses*, 2nd ed., Springer-Verlag, New York.

BERGER–PARKER DOMINANCE INDEX

See also DIVERSITY INDICES; DOMINANCE INDEX, COMMUNITY

Introduction

In ecology, the Berger–Parker dominance index expresses the proportion of total individuals that is due to the dominant species. It is simpler than the dominance index.

Formula

$$d = \frac{N_{\max}}{N_T}$$

where

$N_{\max}$ = abundance of the dominant species
N_T = total number of individuals in the sample

Reference

Southwood, T. R. E. 1978. *Ecological Methods*, Chapman & Hall, London.

BET ISOTHERM

Introduction

The BET (Brunauer, Emmett, Teller) isotherm describes the partitioning of a compound between liquid and solid compartments or phases. This isotherm assumes the adsorption of multilayers of *adsorbate* on the surface of the *adsorbent* (Benefield et al., 1982; Brunauer et al., 1938; Weber, 1972).

Formula

$$q_e = \frac{BCQ^0}{(C_s - C)[1 + (B - 1)(C/C_s)]}$$

where

q_e = amount of adsorbate adsorbed per unit weight of adsorbent at concentration C

B = constant related to the energy of interaction with the surface

C = equilibrium concentration of adsorbate in solution (mg/L or mol/L)

Q^0 = number of moles of solute per unit weight of absorbent to form a complete monolayer on the surface

C_s = saturation concentration of the adsorbate (mg/L or mol/L)

A typical BET isotherm is shown in Figure B4a. A linear form of the BET isotherm is given by the following equation:

$$\frac{C}{(C_s - C)q_e} = \frac{1}{BQ^0} + \frac{B - 1}{BQ^0}\frac{C}{C_s}$$

Figure B4b shows a straight line with a slope $(B - 1)/BQ^0$ and a Y-intercept of $1/BQ^0$.

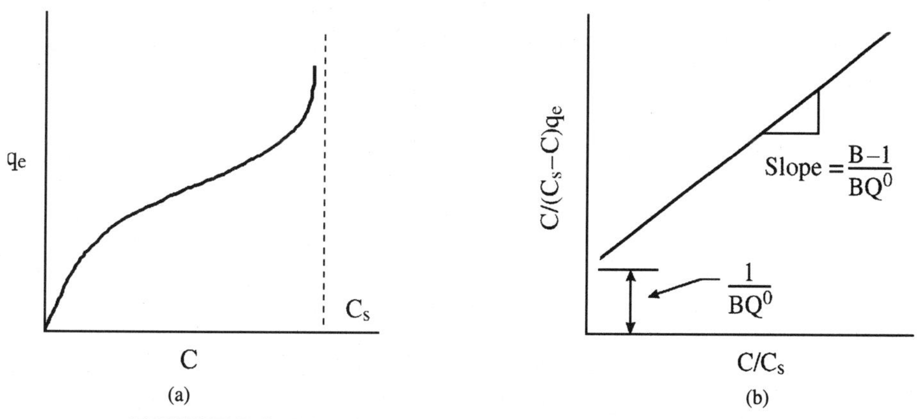

FIGURE B4: BET isotherms: (a) typical; (b) Linear BET.

References

Benefield, L. D., J. F. Judkins, Jr., and B. L. Weand, 1982. *Process Chemistry for Water and Wastewater Treatment*, Prentice Hall, Upper Saddle River, NJ.

Brunauner, S., P. H. Emmett, and E. Teller. 1938. Adsorption of gases in multimolecular layers. J. Am. Chem. Soc. 60: 309–319.

Weber, W. J. Jr. 1972. *Physicochemical Processes for Water Quality Control*, Wiley-Interscience, New York.

BETA (β)

Definition

β is the ratio of the saturation concentration of dissolved oxygen in wastewater to the saturation concentration in clean water.

Formula

$$\beta = \frac{C_s \text{ in wastewater}}{C_s \text{ in clean water}}$$

where C_s is the saturation concentration of dissolved oxygen (mg/L).

Numerical Value

If the value for β is not known, it is generally assumed to be approximately 0.9.

Reference

Institute of Water Pollution Control. 1987. *Unit Processes: Activated Sludge*. IWPC, Maidstone, Kent, England.

BIOCHEMICAL OXYGEN DEMAND (BOD)

Definition/Introduction

BOD is the amount of dissolved oxygen (DO) used by microorganisms in the biochemical oxidation of organic and inorganic matter. It is measured by incubating a sample at a standard temperature (usually 20°C) for a set period of time (commonly 5 days). The sample must be diluted if the BOD exceeds 8 mg/L.

Formula

If the dilution water is not seeded,

$$\text{BOD (mg/L)} = \frac{D_1 - D_2}{P}$$

where

D_1 = initial DO of the sample dilution (mg/L)
D_2 = final DO of the sample dilution (mg/L)
P = decimal volumetric fraction of sample used

Numerical Values

- *Typical BOD_5 values in domestic wastewater*. BOD_5 varies between 110 and 400 mg/L (Metcalf and Eddy, 1979).
- *Correlation with total organic carbon (TOC) and chemical oxygen demand (COD) in untreated domestic wastewater*: For typical untreated domestic wastewater, the BOD_5/COD ratio varies from 0.4 to 0.8, and the BOD_5/TOC ratio varies from 1.0 to 1.6.

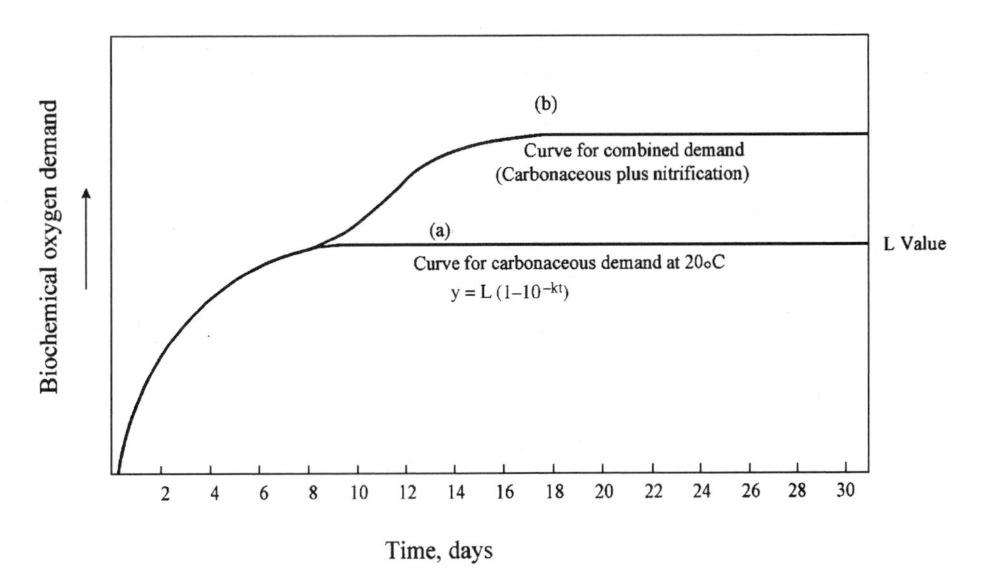

FIGURE B5: Biochemical oxygen demand (BOD): (a) oxidation of organic matter; (b) influence of nitrification.

Carbonaceous Biochemical Oxygen Demand (CBOD) CBOD is the amount of dissolved oxygen used by microorganisms in the biochemical oxidation of organic matter only. It is measured by preventing nitrification during the incubation period of the BOD test. Methods for preventing nitrification include pretreating the sample by adding inhibitory agents such as ammonia, methylene blue, thiourea, allylthiourea, 2-chlor-6 (trichloromethyl) pyridine (TCMP), or proprietary products. Figure B5 shows the influence of nitrification on BOD.

References

Metcalf and Eddy, Inc. 1979. *Wastewater Engineering: Treatment, Disposal and Reuse*, 2nd ed., McGraw-Hill, New York.

Sawyer, C. N., and P. L. McCarty. 1978. *Chemistry for Environmental Engineering*, McGraw-Hill, New York.

Young, J. C. 1973. Chemical methods for nitrification control. J. Water Pollut. Control Fed. 45: 637–646.

BIOCHEMICAL OXYGEN DEMAND CURVE

Introduction

The rate of oxygen consumption by bacteria is proportional to the concentration of organic matter at any time. As shown in Figure B6, as regards the biodegradation of organic matter, the rate of oxygen consumption is rapid in the first few days, then decreases as the concentration of organic matter decreases.

Formula

The rate of BOD decay may be approximated by the expression

$$\mathrm{BOD}_t = L(1 - 10^{-kt})$$

where

t = time (days)

BOD_t = oxygen consumption (BOD) at any time t (mg/L)

L = total or ultimate BOD (= maximum oxygen consumption when the waste has been completely biodegraded) (mg/L)

k = BOD rate constant (day^{-1})

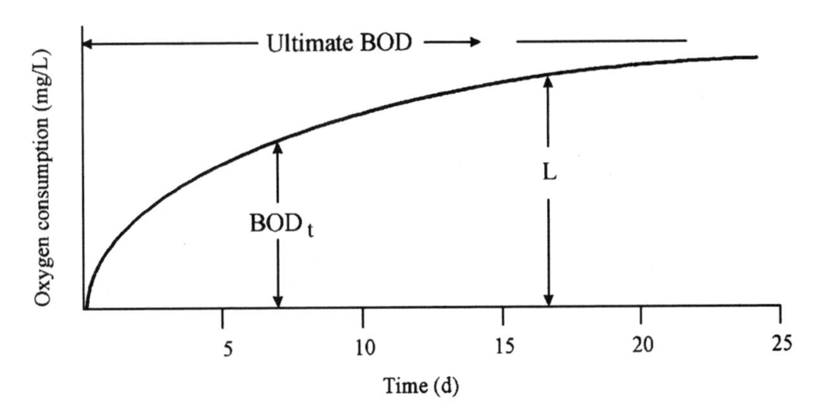

FIGURE B6: BOD curve: Consumption of oxygen during organic matter degradation.

Typical values of the rate constant k, at 20°C, in wastewater, wastewater effluent, and contaminated river water are 0.15–0.30, 0.05–0.10, and 0.05–0.10 day^{-1} respectively (Davis and Cornwell, 1985). The numerical value of k depends on three factors:

- Nature of the wastewater (i.e., types of organic compounds present in the wastewater)
- Ability of the microorganisms to utilize the waste
- Temperature

Since the BOD test is carried out at 20°C, the BOD rate constant k is adjusted to the temperature of the receiving water using the following equation:

$$k_T = k_{20}\theta^{T-20}$$

where

T = temperature (°C)

k_T = BOD rate constant at the temperature of the receiving water (day^{-1})

k_{20} = BOD rate constant at 20°C (day^{-1})

θ = temperature coefficient ($\theta = 1.135$ for temperatures between 4 and 20°C; $\theta = 1.056$ for temperatures between 20 and 30°C)

References

Davis, M. L., and D. A. Cornwell. 1985. *Introduction to Environmental Engineering*, PWS, Boston.

Sawyer, C. N., and P. L. McCarty. 1978. *Chemistry for Environmental Engineering*, McGraw-Hill, New York.

BIOCONCENTRATION FACTOR (BCF)

Definition/Introduction

In toxicology, the bioconcentration factor expresses the bioaccumulation of hydrophobic compounds that tend to assumulate in the fat of animals. It is the ratio of the concentration of a chemical in an organism to that in water. Chemicals with a high partition coefficient K_{ow} (measured in an octanol–water mixture) are greatly prone to bioaccumulation in organisms. There is a good relationship between the bioconcentration factor and the octanol–water partition coefficient.

Formula

$$\text{BCF} = \frac{C_a}{C_w}$$

where

BCF = bioconcentration factor
C_a = toxicant concentration in an organism (μg/g)
C_w = toxicant concentration in water (μg/mL)

Numerical Values

See Table B2.

TABLE B2: Bioconcentrations Factors for Some Hydrophobic Compounds

Compound	Organism	BCF	Concentration in Water
Aldrin	Catfish	1,590	0.044 μg/L
	Buffalo fish	30,000	0.007 μg/L
Chlordane	Algae	302	6.6 ng/L
DDT	Algae	500	0.016 ng/L
	Oyster	70,000	0.1 μg/L
	Trout	200	20 μg/L
Dieldrin	Trout	3,300	2.3 μg/L
PCB	Yellow perch	17,000	1.0 μg/L

Source: Adapted from Jorgensen and Johnsen (1989).

References

Freed, V. H., C. T. Chiou, and R. Hague. 1977. Chemodynamics: transport and behavior of chemicals in the environment—a problem in environmental health. Environ. Health Perspect. 20: 55–70.

Jorgensen, S. E., and I. Johnsen. 1989. *Principles of Environmental Science and Technology*, Elsevier, Amsterdam.

Landrum, P. F., H. Lee II, and M. J. Lidy. 1992. Toxicokinetics in aquatic systems: model comparisons and use in hazard assessment. Environ. Toxicol. Chem. 11: 1709–1725.

Rand, G. M., and S. R. Petrocelli, Eds. 1985. *Fundamentals of Aquatic Toxicology*, Hemisphere, Washington, DC.

BIOFILM GROWTH KINETICS

Introduction

Adsorption of microorganisms to solid surfaces results in the formation of biofilms.

Formula

Microbial growth in biofilms is described by the following equation (La Motta, 1976; Uhlmann, 1979):

$$\frac{dX}{dt} = \mu X - k_d X$$

where

μX = biomass growth
$k_d X$ = biomass loss
X = number of microorganisms
μ = specific growth rate (h^{-1})
k_d = decay rate (h^{-1})

The growth rate μ depends on the wastewater loading rate, while k_d depends on factors such as removal by grazing, decay of bacterial and fungal biomass, and hydraulic loading. At steady state, $dX/dT = 0$, so $\mu X = k_d X$.

NEW YORK INSTITUTE OF TECHNOLOGY LIBRARY

References

La Motta, E. J. 1976. Internal diffusion and reaction in biological films. Environ. Sci. Technol. 19: 765–769.

Uhlmann, D. 1979. *Hydrobiology: A Text for Engineers and Scientists*, Wiley, New York.

BIOFILMS: Steady-state model

Introduction

The steady-state model (microbial biomass growth balances biomass loss) describes the growth of biofilm bacteria in the presence of a single growth-limiting substrate (Rittmann and McCarty, 1980; McCarty et al., 1986). The model considers the diffusion of the substrate from the bulk solution into the biofilm and its subsequent utilization by biofilm bacteria, as well as bacterial growth and decay (Figure B7).

Formula

$$L_f = \frac{YJ}{bX_f}$$

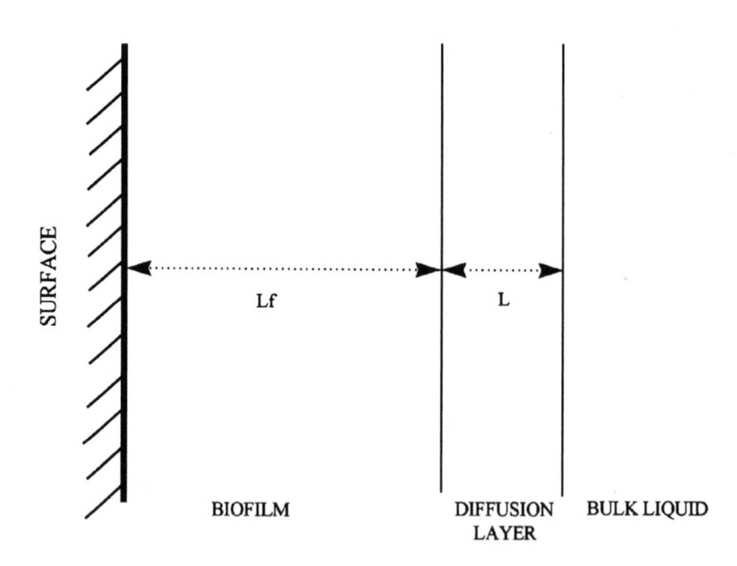

FIGURE B7: Biofilm.

where

L_f = steady-state biofilm thickness (cm)

Y = true yield of cell biomass per substrate used (mg VSS/mg substrate)

J = substrate flux ($mg/cm^2 \cdot day$). It is the rate at which a substrate is transported from the bulk solution into the biofilm (Rittmann, 1987).

b = bacterial decay rate (day^{-1})

X_f = biofilm cell density (mg VSS/cm^3)

J is given by Fick's first law:

$$J = D\frac{S - S_s}{L}$$

where

D = molecular diffusivity of substrate in water (cm^3/day)

S = substrate concentration in the bulk solution (mg/L)

S_s = substrate concentration at the biofilm surface (mg/L)

L = effective diffusion layer (cm)

References

McCarty, P. L, B. E. Rittmann, and E. J. Bouwer. 1986. Microbiological processes affecting chemical transformations in groundwater, pp. 89–115. In: *Groundwater Pollution Microbiology*, G. Bitton and C. P. Gerba, Eds., Wiley, New York.

Rittmann, B. E. 1987. Aerobic biological treatment. Environ. Sci. Technol. 21: 128–136.

Rittmann, B. E., and P. L. McCarty. 1980. Model of steady-state biofilm kinetics. Biotechnol. Bioeng. 22: 2243–2357.

BIOLOGICAL HALF-LIFE ($t_{1/2}$)

Definition

Biological half-life is the time it takes for a compound to decrease to half of its initial concentration.

Formula

$$t_{1/2} = \frac{\ln 2}{k} = \frac{0.693}{k}$$

where k is the first-order decay rate (1/time).

BIOMASS, BACTERIAL

See also ADENOSINE TRIPHOSPHATE (ATP); BIOVOLUME MICROBIAL; LIPOPOLYSACCARIDE (LPS); MURAMIC ACID.

Introduction

Bacterial biomass may be obtained in the following manner: Size estimates of bacteria are obtained via scanning electron microscopy or via epifluorescence microscopy. The biovolumes of rods (considered as cylinders) and cocci (considered as spheres) are computed. Volumes are converted to biomass by multiplying by 1.21×10^{-13} g C/μm^3 (Fuhrman and Azam, 1980; Watson et al., 1977).

Numerical Values

Some estimates of average biomass per cell in coastal waters are listed in Table B3.

References

Fuhrman, J. A., and F. Azam, 1980. Bacterioplankton secondary production estimates for coastal waters of British Columbia, Antarctica and California. Appl. Environ. Microbiol. 39: 1085–1095.

Watson, S. W., T. J. Novitsky, H. L. Quinby, and F. W. Valois. 1977. Determination of bacterial numbers and biomass in the marine environment. Appl. Environ. Microbiol. 33: 940–946.

TABLE B3: Biomass of Microorganisms in Coastal Waters

	Microbial Biomass (fg C/cell)	
Location	Average	Range
Saanich Inlet (Canada)	13	9.7–17
Antarctica	8.3	7.0–9.1
California	20	4.6–39

Source: Adapted from Fuhrman and Azam (1980).

BIOVOLUME, MICROBIAL

See also BIOMASS, BACTERIAL.

Definition/Introduction

Biovolume, the volume occupied by a microbial cell, has been used to estimate microbial biomass. Cell volume can be measured using flow cytometry or microscopic techniques (epifluorescence microscopy, scanning electron microscopy, transmission electron microscopy). Microscopic techniques, although straightforward, do not, however, distinguish between dead and live microorganisms. Methods for determining cell volumes are described by Bratbak (1993).

Formula (Bratback and Dundas, 1984)

The microbial cells can be fixed with glutaraldehyde and their width (W) and length (L) are measured. Cell volumes (V) are then determined according to the following formulas:

Rods

$$V = \frac{\pi}{4} W^2 \left(L - \frac{W}{3} \right)$$

According to Bratbak (1993), the formula for rods above works equally well for cocci

Cocci (Lee, 1993)

$$V = \tfrac{4}{3} \pi R^3$$

- According to Posch et al. (1997) bacteria are considered as cylinders with two spherical ends:

$$V = \left(\frac{W^3}{6} \times \pi \right) + (L - W) \times \left(\frac{W^2}{4} \times \pi \right)$$

where

V = volume (μm^3)

W = microbial cell width (μm)

L = microbial cell length (μm)

R = sphere radius (μm)

Numerical values

Bacteria

- *Bacterial cultures.* See Table B4. For a continuous mixed culture of soil bacteria (*Arthrobacter* sp., *Alcaligenes* sp., and *Corynebacterium michiganense*), cell volume varied between 0.25 and 0.41 μm^3 (Bloem et al., 1995).
- *Environmental samples.* Bacterial volumes from a wide range of aquatic environments are shown in Table B5. In a granular activated carbon (GAC) filter in a water treatment plant, the mean biovolume of bacteria was 0.18 μm^3 (Servais et al., 1991).
- *Effect of starvation on bacterial biovolume.* Starvation tends to reduce bacterial size. Bacteria respond to starvation by forming ultramicrocells (Kjellerberg et al., 1993). For example, actively growing *Vibrio* cells have a biovolume of 5.94 μm^3, while the biovolume of starved cells may be as low as 0.5 μm^3 (Moyer and Morita, 1989).

Protozoa The estimated cell volumes of protozoan ciliates collected from freshwater lakes in Florida was reported to vary between 900 and 150,000 μm^3 (Table B6).

Algae The biovolumes of some freshwater algae are shown in Table B7.

Biovolume-to-Carbon Conversion

- *Bacteria.* The conversion factor (fg C/μm^3) ranges between 75 and 560 (Bowden, 1977; Bratbak, 1985; Hagstrom et al., 1979; Nagata, 1986; Watson et al., 1977). Fry (1990) observed that conversion factors obtained using electron microscopy are generally lower than those obtained by other

TABLE B4: Biovolume of *P. putida* Cells According to Culture Conditions

	Biovolume (μm^3)		
		Epifluorescence Microscopy	
Limiting Nutrient	SEM	Eyepiece Graticule	Photo
C	0.30	0.28	0.29
N	0.39	0.66	0.71
P	0.34	0.57	0.63

Source: Adapted from Bratbak (1985).

TABLE B5: Estimates of Mean Cell Volumes in Aquatic Habitats

Type of Bacteria	Range of Mean Cell Volumes (μm^3)
Planktonic	
Marine	0.036–0.19
Estuarine	0.047–0.096
Freshwater	0.005–0.77
Sediment	0.088–0.98
Epiphytic	0.31–0.49
Epilithic	0.10–0.40

Source: Data from Bowden (1977), Fry (1988), Morris and Lewis (1992), Riemann (1983), Scavia and Laird (1987), Tumber et al. (1994), Vadstein et al. (1989).

TABLE B6: Biovolumes of Some Ciliates from Florida Lakes

Ciliate	Biovolume (μm^3)
Haptorida	
Cyclotrichium	900
Spathidium	3,000
Didinium	40,293
Oligotrichida	
Strombidium	20,000
Tintinnopsis	100,000
Tintinnidium	150,000
Scuticociliatida	
Cyclidium	1,125–3,381
Uronema	3,380
Prostomatida	
Urotricha	2,048
Holophora	4,096
Trichostomatida	
Plagtopyla	96,000
Pleutostomatida	
Litonotus	24,000
Peritrichida	
Vorticella	13,500
Hymenstomatida	
Paramecium	120,000

Source: Adapted from Beaver and Crisman (1982).

TABLE B7: Calculated Mean Cell Volumes of Representative Species of Freshwater and Marine Organisms

Microorganism	Biovolume (μm^3)	Microorganism	Biovolume (μm^3)
Cyanophyta		Bacillariophyceae	
Anabaena flos-aquae (col.)	80,000	*Asterionella formosa* (Michigan)	350
Aphanocapsa delicatissima	4	*Diatoma vulgare*	4,350
Chroococcus limneticus (col.)	400	*Melosira islandica* (1 mm)	80,000
Microcystis aeruginosa (col.)	100,000	*Nitzchia gracilis*	240
Microcystis flos-aquae	50	*Synedra acus*	250
Oscillatoria limnetica (1 mm)	17,500	*Tabellaria fenestrata* (Michigan)	3,000
Synechococcus aeruginosus	350	*Thalassiosira oceanica* (open-ocean diatom)	75
		Thalassiosira weissflogii (coastal diatom)	800
Chlorophyta		Pyrrophyta	
Ankistrodesmus falcatus	250	*Ceratium hirundinella*	4,000
Botryococcus braunii	10,000	*Gymnodinium fuscum*	10,000
Chlamydomonas subcompleta	250	*Heterocapsa pygmaea* (marine)	600
Chlorella vulgaris	200	*Perdinium cinctum*	40,000
Cosmarium reniforme	30,000	Euglenophyta	
Dunaliella tertiolecta (marine)	300	*Trachelomonas hispida*	4,200
Pandorina morum	4,000	*Trachelomonas volvocina*	1,800
Scenedesmus quadricauda	1,000		
Staurastrum paradoxum	20,000		
Tetraselmis maculata (marine)	425		
Chrysophyta			
Dinobryon borgei	1,500		
Mallomonas caudata	12,000		
Rhizochrysis limnetica	1,200		

Source: Data from Wetzel (1983), Ahner et al. (1995).

TABLE B8: Conversion Factors of Bacterial Biovolumes to Nitrogen Biomass

Source[a]	Mean Cell Volume (μm^3)	Conversion Factor (mean pg N/μm^3)	Reference
A	0.376	104 (45–202)	Bratbak (1985)
B	0.163	25 (NS–41)	Nagata (1986)
	0.545	27 (7–42)	Nagata and Watanabe (1990)
C	0.290	30 (11–50)	Kogure and Koike (1987)
	0.056	110 (60–220)	Lee and Fuhrman (1987)

Source: Adapted from Lee (1993).

[a] A, pure culture/seawater; B, freshwater; C, seawater.

[b] NS, nonsignificant number.

methods because of cell shrinkage. Fry (1988, 1990) recommended a conversion factor of 310 fg C/μm^3. Bratbak (1993) suggested a conversion factor of about 350 fg C/μm^3 and a value of 200 fg C/μm^3 if a more conservative factor is sought. Simon and Azam (1989) suggested a conversion factor of 207 fg C/μm^3. Watson et al. (1977) recommended a conversion factor of 121 fg C/μm^3. In Lake Michigan the conversion factor found was 154 ± 47 fg C/μm^3 (Scavia and Laird, 1987), whereas Nagata (1986) derived a factor of 106 fg/μm^3. Bloem et al. (1995), using confocal laser scanning microscopy and image analysis, reported a value of 200 fg C/μm^3 for continuous mixed cultures of soil bacteria (*Arthrobacter* sp., *Alcaligenes* sp., and *Corynebacterium michiganense).*

- *Protozoa.* Heterotrophic flagellates (*Monas* sp.) (Borsheim and Bratbak, 1987):

  Living flagellate: 100 fg C/μm^3

  Preserved flagellate: 220 C/μm^3
- *Algae.* Phytoplankton in eutrophic lake in Norway: 0.14 fg C/μm^3 (Vadstein et al., 1989).

Biovolume-to-Nitrogen Conversion Conversion factors are given in Table B8.

References

Ahner, B. A., S. Kong, and F. M. M. Morel. 1995. Phytochelatin production in marine algae. 1. An interspecies comparison. Limnol. Oceanogr. 40: 649–657.

Beaver, J. R., and T. L. Crisman. 1982. The trophic response of ciliated protozoans in freshwater lakes. Limnol. Oceanogr. 27: 246–253.

Bloem, J., M. Veninga, and J. Shepherd. 1995. Fully automatic determination of soil bacterium numbers, cell volumes, and frequency of dividing cells by confocal laser scanning microscopy and image analysis. Appl. Environ. Microbiol. 61: 926–936.

Borsheim, K. Y., and G. Bratbak. 1987. Cell volume to cell carbon conversion factors for a bacterivorous *Monas* sp. enriched from seawater. Mar. Ecol. Prog. Ser. 36: 171–175.

Bowden, W. B. 1977. Comparison of two direct-count techniques for enumerating aquatic bacteria. Appl. Environ. Microbiol. 33: 1229–1232.

Bratbak, G. 1985. Bacterial biovolume and biomass estimations. Appl. Environ. Microbiol. 49: 1488–1493.

Bratbak, G. 1993. Microscope methods for measuring bacterial biovolume: epifluorescence microscopy, scanning electron microscopy and transmission electron microscopy, pp. 309–317. In: *Handbook of Methods in Aquatic Microbial Ecology*, O. F. Kemp, B. F. Sherr, E. B. Sherr, and J. J. Coles, Eds., Lewis Publishers, Bota Raton, FL.

Bratback, G., and I. Dundas. 1984. Bacterial dry matter content and biomass estimations. Appl. Environ. Microbiol. 48: 755–757.

Fry, J. C. 1988. Determination of biomass, pp. 27–72. In: *Methods in Aquatic Bacteriology*, B. Austin, Ed., Wiley, Chichester, West Sussex, England.

Fry, J. C. 1990. Direct methods and biomass estimation. Methods Microbiol. 22: 43–85 (R. Grigorova and J. R. Norris, Eds.), Academic Press, London.

Gaedke, U. 1992. The size distribution of plankton biomass in a large lake and its seasonal variability. Limnol. Oceanogr. 37: 1202–1220.

Hagstrom, A., U. Larsson, P. Horsted, and S. Normark. 1979. Frequency of dividing cells, a new approach to the determination of bacterial growth rates in aquatic environments. Appl. Environ. Microbiol. 37: 805–812.

Kjelleberg, S., K. B. G. Flärdh, T. Nyström, and D. J. W. Moriarty. 1993. Growth limitation and starvation of bacteria, pp. 289–320. In: *Aquatic Microbiology: An Ecological Approach*, T. E. Ford, Editor, Blackwell Sci. Pub., Boston, U.S.A.

Kogure, K., and I. Koike. 1987. Particle counter determination of bacterial biomass in seawater. Appl. Environ. Microbiol. 53: 274.

Lee, S. 1993. Measurement of carbon and nitrogen biomass and biovolume from naturally derived marine bacterioplankton, pp. 319–325. In: *Handbook of Methods in Aquatic Microbial Ecology*, P. F. Kemp, B. F. Sherr, E. B. Sherr, and J. J. Cole, Eds., Lewis Publishers, Boca Raton, FL.

Lee, S., and J. A. Fuhrman. 1987. Relationships between biovolume and biomass of naturally derived bacterioplankton. Appl. Environ. Microbiol. 53: 1298–1305.

Linley, E. A. S., R. C. Newell, and M. I. Lucas. 1983. Quantitative relationships between phytoplankton, bacteria and heterotrophic microflagellates in shelf waters. Mar. Ecol. Prog. Ser. 12: 77–89.

Morris, D. P., and W. M. Lewis, Jr. 1992. Nutrient limitation of bacterioplankton growth in Lake Dillon, Colorado. Limnol. Oceanogr. 37: 1179–1192.

Moyer, C. L., and R. Y. Morita. 1989. Effect of growth rate and starvation-survival on the viability and stability of a psychrophilic marine bacterium. Appl. Environ. Microbiol. 55: 1122–1127.

Nagata, T. 1986. Carbon and nitrogen content of natural planktonic bacteria. Appl. Environ. Microbiol. 52: 28–32.

Nagata, T., and Y. Watanabe. 1990. Carbon- and nitrogen-to-volume ratios of bacterioplankton grown under different nutritional conditions. Appl. Environ. Microbiol. 56: 1303–1309.

Porsch, T., J. Pernthaler, A. Alfreider, and R. Psenner. 1997. Cell-specific respiratory activity of aquatic bacteria studied with the tetrazolium reduction method, cyto-clear slides, and image analysis. Appl. Environ. Microbiol. 63: 867–873.

Riemann, B. 1983. Biomass and production of phyto- and bacterio-plankton in eutrophic Lake Tystrup, Denmark. Freshwater Biol. 13: 389–398.

Scavia, D. and G. A. Laird. 1987. Bacterioplankton in Lake Michigan: dynamics, controls, and significance to carbon flux. Limnol. Oceanogr. 32: 1017–1033.

Servais, P., G. Billen, C. Ventresque, and G. P. Balbon. 1991. Microbiol activity in GAC filters at the Choisy-le-Roi treatment plant. J. Am. Water Works Assoc. 83: 62–68.

Simon, M., and F. Azam. 1989. Protein content and protein synthesis rates of planktonic marine bacteria. Mar. Ecol. Prog. Ser. 51: 201–213.

Tumber, V. P., R. D. Roberts, M. T. Arts, M. S. Evans, and D. E. Caldwell. 1994. Influence of environmental factors on seasonal changes in bacterial cell volumes in two prairie saline lakes. Microb. Ecol. 26: 9–20.

Vadstein, O., B. O. Harkjerr, A. Jensen, Y. Olsen, and C. Reinertsen. 1989. Cycling of organic carbon in the photic zone of a eurotrophic lake with special reference to the heterotrophic bacteria. Limnol. Oceanogr. 34: 840–855.

Watson, S. W., T. J. Novitsky, H. L. Quinby, and F. W. Valois. 1977. Determination of bacterial number and biomass in the marine environment. Appl. Environ. Microbiol. 33: 940–946.

Wetzel, R. G. 1983. *Limnology*, 2nd ed., Saunders College Publishing, Philadelphia.

BIRTH RATES (*DAPHNIA*)

Formula

Birth rate b of *Daphnia* is determined according to the following equation:

$$b = \frac{\ln(1 + E/N)}{D}$$

where

b = birthrate (day^{-1})
E = total number of eggs
N = total number of females
E/N = the egg ratio
D = development time of eggs (days)

$D = 2.8$ days at 20°C (Elster and Schworbel, 1970).

References

Elster, H. J., and J. Schworbel. 1970. Beiträge zur Biologie and Populationdynamik der Daphnien in Bodensee. Arch. Hydrobiol. Suppl. 38: 18–72.

Paloheimo, J. H. 1974. Calculations of intantaneous birth rate. Limnol. Oceanogr. 1: 692–694.

Sommer, U. 1993. Phosphorus-limited *Daphnia*: intraspecific facilitation instead of competition. Limnol. Oceanogr. 37: 966–973.

BOLTZMANN CONSTANT

Definition

The Boltzmann constant, the gas constant per molecule, is the ratio of the gas constant to the Avogadro number.

Formula

$$k = \frac{R}{N} = 1.380658 \times 10^{-23} \text{ J/K}$$

where

k = Boltzman constant
R = gas constant = 8.314510 J/mol · K
N = Avogadro's number = 6.0221367×10^{23} mol^{-1}

Reference

Williams, W. R., W. L. Mattice, and H. B. Williams. 1978. *Basic Physical Chemistry for the Life Sciences*, W.H. Freeman, San Francisco.

BOLTZMANN EQUATION

Introduction

The Boltzmann equation describes the distribution of counterions at a charged surface, based on the Gouy–Chapman diffuse double-layer theory (*see also* ELECTRICAL DOUBLE LAYER).

Formula

$$n(x) = n_0 e^{-ze\Psi(x)/kT}$$

where

$n(x)$ = concentration of counterions at a distance x from the charged surface
n_0 = concentration of counterions in the bulk solution
e = unit of electronic charge
z = charge of the counterion
$\Psi(x)$ = electrical potential at a distance x from the surface
k = Boltzmann constant
T = absolute temperature (K)

Reference

McBride, M. B. 1994. *Environmental Chemistry of Soils*, Oxford University Press, New York.

BROWNIAN MOTION

Definition

Brownian motion is the random motion of colloidal particles in liquids caused by the collisions between molecules and suspended particles.

Formula

The Brownian displacement (Δ) of a particle, including microorganisms, in a given direction is given by

$$\Delta = 2Dt$$

where

D = diffusion coefficient (cm^2/s)
t = time (s)

TABLE B9: Brownian Displacements and Diffusion Coefficients Calculated for Uncharged Spheres in Water at 20°C

Particle Radius	Diffusion Coefficient (cm^2/s)	Brownian Displacement (cm/h)
10^{-9} m (1 nm)	2.1×10^{-6}	1.23×10^{-1}
10^{-8} m (10 nm)	2.1×10^{-7}	3.90×10^{-2}
10^{-7} m (0.1 mm)	2.1×10^{-8}	1.23×10^{-2}
10^{-6} m (1.0 mm)	2.1×10^{-9}	3.90×10^{-3}

Source: Adapted from Shaw (1996).

The diffusion coefficient D of suspended colloidal particles is given by

$$D = \frac{RT}{6\pi\eta a N}$$

where

R = universal gas constant (J/mol · K)
T = absolute temperature (K)
η = viscosity (cP)
a = particle radius (mm)
$N = 6.023 \times 10^{23}$

Numerical Values

See Table B9.

References

Marshall, K. C. 1976. *Interfaces in Microbial Ecology*, Harvard University Press, Cambridge, MA.

Shaw, D. J. 1966. *Introduction to Colloid and Surface Chemistry*, Butterworth, London.

BULK DENSITY OF SOILS

Definition

Bulk density is the ratio of the mass of solids of a given quantity of soil to the total volume of the soil.

TABLE B10: Bulk Densities of Soils

Soil Type	Bulk Density (g/cm^3)
Sand	1.55
Sandy loam	1.40
Fine sandy loam	1.30
Loam	1.20
Clay loam	1.10
Clay	1.05

Source: Adapted from Hausenbuiller (1978).

Formula

$$\rho_b = \frac{m_s}{V_t}$$

where

ρ_b = soil bulk density (g/cm^3)
m_s = mass of solids (g)
V_t = total volume of soil (cm^3)

Numerical Values

(See Table B10. According to Donahue et al. (1977), bulk density suitable for plant growth should be 1.4 and 1.6 g/cm^3 for clays and sands, respectively.

References

Donahue, R. L., R. W. Miller, and J. C. Shikluma. 1977. *Soils: An Introduction to Soils and Plant Growth*, Prentice-Hall, Upper Saddle River, NJ.

Hausenbuiller, R. L. 1978. *Soil Science: Principles and Practice*, Wm. C. Brown, Dubuque, IA.

Marshall, T. J., and J. W. Holmes. 1988. *Soil Physics*, 2nd ed., Cambridge University Press, Cambridge.

C

CARBON CONTENT OF MICROORGANISMS

See also BIOVOLUME, MICROBIAL.

Numerical Values

Bacterial Cells Range: 1.1×10^{-13} to 3.1×10^{-13} g C/cell (Bratbak, 1985). In a granular activated carbon (GAC) filter in a water treatment plant, the carbon content was 2.16×10^{-14} g C/bacterial cell (Servais et al., 1991).

Algae *(e.g., Chlamydomonas reinhardi)* The carbon content of *Chlamydomonas reinhardi* cell was found to be 2.0×10^{-11} g/cell (Porter et al., 1982).

References

Bratbak, G. 1985. Bacterial biovolume and biomass estimations. Appl. Environ. Microbiol. 49: 1488–1493.

Porter, K. G., J. Gerritsen, and J. D. Orcutt, Jr. 1982. The effect of food concentration on swimming patterns, feeding behaviour, ingestion, assimilation, and respiration by *Daphnia*. Limnol. Oceanogr. 27: 935–949.

Servais, P., G. Billen, C. Ventresque, and G. P. Balbon. 1991. Microbial activity in GAC filters at the Choisy-le-Roi treatment plant. J. Am. Water Works Assoc. 83: 62–68.

CARBON CONVERSION EFFICIENCY: Protozoa

Definition/Introduction

Protozoa feed on microbial cells such as bacterial and yeast cells. The carbon conversion efficiency (CeF) is the percent of the prey carbon that is incorporated into protozoan cells.

Formula

$$\text{CeF}\ (\%) = \frac{\text{carbon incorporated into protozoan biomass}}{\text{prey carbon used}}$$

Note: The carbon conversion efficiency determination assumes that the protozoa only ingest bacteria (Turley, 1993).

Numerical Values

Turley (1993) gave literature data on carbon conversion efficiencies.

- Shallow-water flagellates: CeF = 24–62%
- Barophilic flagellates: CeF = 17–25%

Reference

Turley, C. M. 1993. Determination of pressure effects on flagellates isolated from surface water, pp. 91–96. In: *Handbook of Methods in Aquatic Microbial Ecology*, P. F. Kemp, B. F. Sherr, E. B. Sherr, and J. Cole, Eds., Lewis Publishers, Boca Raton, FL.

CARBON/NITROGEN RATIO

See C/N RATIO OF MICROORGANISMS.

CARBON WEIGHT OF BACTERIAL CELLS: Relationship with Cell Wet and Dry Weights

Formulas and Numerical Values

Carbon/Wet Biomass Ratio

$$\frac{\text{Carbon weight}}{\text{Wet biomass}} = 0.079\text{–}0.129$$

This ratio varies between 0.079 for bacterial populations in natural waters (Fergusson and Rublee, 1976; Bowden, 1977) and 0.129 for cultured bacteria (Watson et al., 1977; Krambeck et al., 1981). A ratio of 0.15 was used to calculate bacterial abundance in coastal waters in Sweden (Hagstrom et al., 1979). A mean ratio of 0.1 is generally used (Troitsky and Sorokin, 1967; Linley et al., 1983).

Carbon/Dry Weight Ratio (Bacterial) (Bratbak and Dundas, 1984; Luria, 1960; Larsson and Hagstrom, 1982; Nagata, 1986; Watson et al., 1977)

$$\frac{\text{Carbon weight}}{\text{Dry weight}} = 0.5$$

Carbon/Dry Weight Ratio (Algal) (Wetzel, 1983)

$$\frac{\text{Carbon weight}}{\text{Dry weight}} = 0.53 \pm 0.05$$

References

Bowden, W. B. 1977. Comparison of two direct count techniques for enumerating aquatic bacteria. Appl. Environ. Microbiol. 33: 1229–1232.

Bratbak, G., and I. Dundas. 1984. Bacterial dry matter content and biomass estimations. Appl. Environ. Microbiol. 48: 755–757.

Ferguson, R. L., and P. Rublee. 1976. Contribution of bacteria to the standing crop of coastal plankton. Limnol. Oceanogr. 21: 141–145.

Hagstrom, A., U. Larsson, P. Horsted, and S. Normark, 1979. Frequency of dividing cells, a new approach to the determination of bacterial growth rates in aquatic environments. Appl. Environ. Microbiol. 37: 805–812.

Krambeck, C., H. J. Krambeck, and J. Overbeck. 1981. Microcomputer-assisted biomass determination of plankton bacteria on scanning electron micrographs. Appl. Environ. Microbiol. 42: 142–149.

Larsson, U., and A. Hagstrom. 1982. Fractionated phytoplankton primary production, exudate release and bacterial production in a Baltic eutrophication gradient. Mar. Biol. 67: 57–70.

Linley, E. A. S., R. C. Newell, and M. I. Lucas. 1983. Quantitative relationships between phytoplankton, bacteria and heterotrophic microflagellates in shelf waters. Mar. Ecol. Prog. Ser. 12: 77–89.

Luria, S. E. 1960. The bacterial protoplasm: composition and organization. pp. 1–34. In: *The Bacteria*, vol. 1, I. C. Gunsalus and R. Y. Stanier, Eds., Academic Press, San Diego, CA.

Nagata, T. 1986. Carbon and nitrogen content of natural planktonic bacteria. Appl. Environ. Microbiol. 52: 28–32.

Troitsky, A. S., and Y. I. Sorokin. 1967. On the methods of calculations of the bacterial biomass in water bodies. Trans. Inst. Biol. Inland Waters Acad. Sci. USSR 19: 85–90.

Watson, S. W., T. J. Novitsky, H. L. Quinby, and F. W. Valois. 1977. Determination of bacterial numbers and biomass in the marine environment. Appl. Environ. Microbiol. 33: 940–946.

Wetzel, R. G. 1983. *Limnology*, 2nd ed., Saunders College Publishing, Philadelphia.

CHEMICAL OXYGEN DEMAND (COD): Relationship with Ultimate BOD

See also BIOCHEMICAL OXYGEN DEMAND CURVE.

Definition/Introduction

Chemical oxygen demand (COD) is the amount of oxygen necessary to oxidize the organic carbon completely to CO_2 and H_2O. Some organic chemicals are not completely oxidized, however.

COD is measured in approximately 3 hours via oxidation with potassium dichromate ($K_2Cr_2O_7$) in the presence of sulfuric acid and silver. During the COD test, other reduced substances (e.g., sulfides, sulfites, Fe^{2+}) are also oxidized and included in the COD. Furthermore, reduced forms of organic nitrogen are converted to ammonia in the COD test and more oxidized forms of nitrogen (e.g., nitrite) are converted to nitrate (Sawyer and McCarty, 1978). If the COD value is much higher than the BOD value, it means that the sample contains large amounts of organic compounds that are not easily biodegraded. For some wastewaters, COD can be correlated with BOD. For example, COD of a 500 mg/L solution of phenol (C_6H_6O):

$$C_6H_6O + 7O_2 \rightarrow 6CO_2 + 3H_2O$$

$$\mathrm{COD} = \frac{7(32)}{94} \times 500\ \mathrm{mg/L} = 1191.5\ \mathrm{mg/L}$$

Relationship with Ultimate BOD For wastewaters with readily degradable organics (e.g., dairy wastes), the COD is given by

$$\mathrm{COD} = \frac{\mathrm{BOD_{ult}}}{0.92}$$

where $\mathrm{BOD_{ult}}$ is the ultimate BOD. (mg/L)

Numerical Values

- *Untreated domestic wastewaters:* COD range = 250–1000 mg/L (Metcalf and Eddy, 1991)
- *Industrial wastewaters:* COD range = 200–350,000 mg/L (Eckenfelder, 1989)

References

Eckenfelder, W. W. Jr., 1989. *Industrial Water Pollution Control,* 2nd ed., McGraw-Hill, New York.

Metcalf and Eddy, Inc. 1991. *Wastewater Engineering: Treatment, Disposal and Reuse,* 3rd ed., McGraw-Hill, New York.

Sawyer, C. N., and P. L. McCarty. 1978. *Chemistry for Environmental Engineering,* 3rd ed., McGraw-Hill, New York.

CHEMOSTAT: Dilution Rate

Definition

In a chemostat for continuous culture of microorganisms, the dilution rate D is the rate at which fresh substrate is distributed throughout the culture vessel.

Formula

$$D = \frac{F}{V}$$

where

D = dilution rate (h^{-1})

F = flow rate (mL/hr)

V = volume of growth medium in the reactor (mL)

The reciprocal of the dilution rate D is called *the mean cell residence time*, which is the time that microorganisms remain in the culture vessel.

Reference

Slater, J. H. 1979. Microbial population and community dynamics, pp. 45–63. In: *Microbial Ecology: A Conceptual Approach*, J. M. Lynch and N. J. Poole, Eds., Blackwell Scientific, Oxford.

CHEMOSTAT: (Microbial Growth Kinetics)

See CONTINUOUS CULTURE OF MICROORGANISMS.

CHICK'S LAW

See also REACTION KINETICS: First-Order Reaction.

Introduction

At the beginning of this century, Harriet Chick postulated that for a given disinfectant and concentration, the death of microorganisms follows first-order kinetics with respect to time. Chick's law can also be applied to express the decrease in titer of a microbial population due to other inactivating factors in an unfavorable environment.

Formula

$$-\frac{dX}{dt} = kX$$

where

X = concentration of living microorganisms at time t

k = first-order decay rate (1/time)

The integrated form of Chick's law is as follows:

$$X = X_0 e^{-kt}$$

or

$$\ln \frac{X}{X_0} = -kt$$

where

X = concentration of living microorganisms at time t (number/unit volume)

X_0 = initial concentration of living microorganisms (number/unit volume)

k = decay rate (1/time)

t = time

Chick's law is represented graphically as a straight line when plotting $\log_{10}$ or $\ln(X/X_0)$ versus time t (Figure C1). The plot has a slope of $-k$. Some plot $-\ln(X/X_0)$ versus t and obtain a straight line with an intercept of zero and a slope k (Ray, 1995).

Numerous laboratory and field studies have shown that the persistence of pathogenic and nonpathogenic microorganisms in ponds, streams, and other aquatic environments approximated Chick's law.

As regards disinfection, Chick's law assumes a constant concentration of disinfectant, uniform susceptibility of all microbial species present, and the absence of interfering substances (Weber, 1972). However, field inactivation data actually show a deviation from first-order kinetics (Figure C2). Curve B in Figure C2 shows deviation from first-order kinetics. The tailing off of the curve results from the survival of a resistant subpopulation within a heterogeneous population, or from protection of the pathogens by interfering factors (e.g., inorganic and organic compounds, turbidity). Microbial clumping explains the "shoulder" of

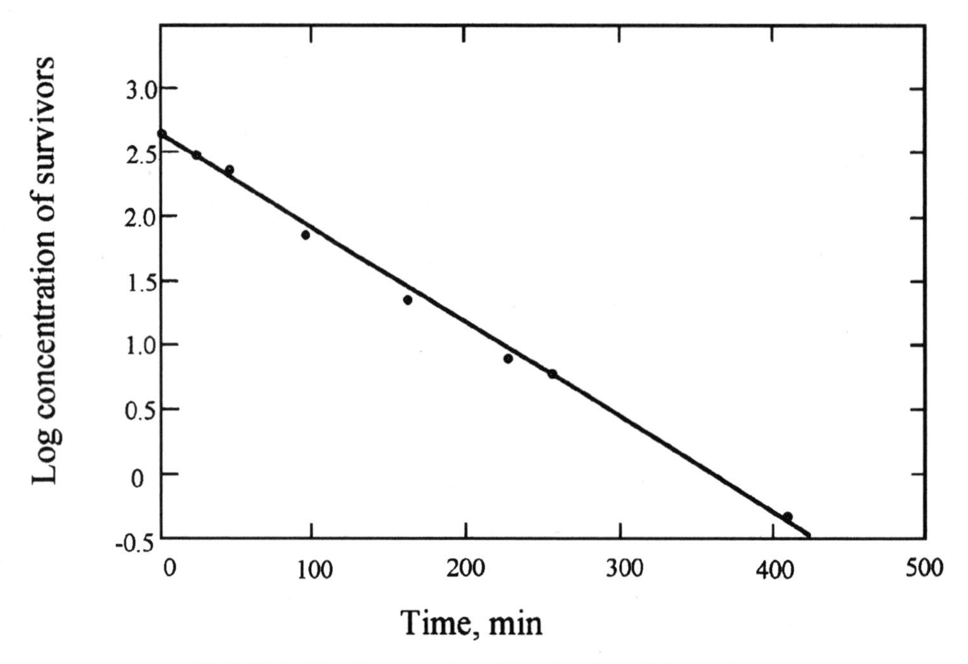

FIGURE C1: First-Order Kinetics for Chick's Law.

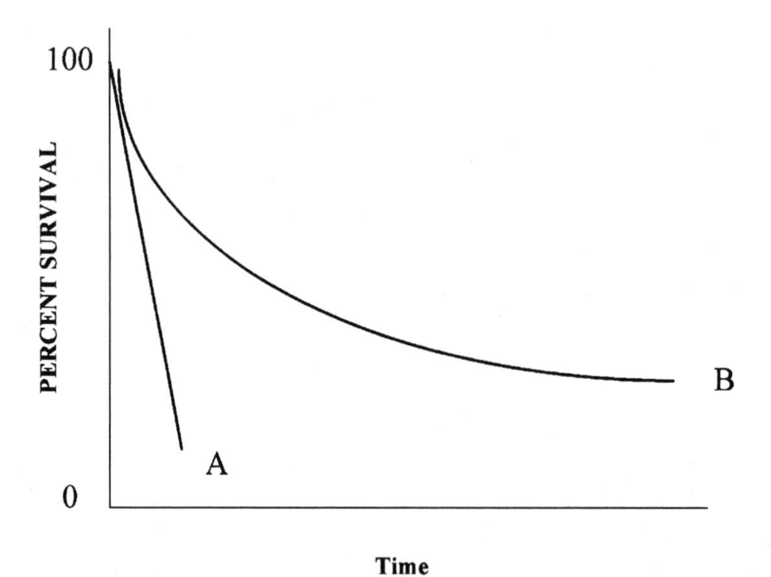

FIGURE C2: Disinfection: Deviation from First-Order Kinetics (*Source:* adapted from Hoff and Akin, 1986.)

survival curves obtained when exposing microorganisms to chlorine action (Rubin et al., 1983).

There are several modifying factors that affect the decay rate constant k. The most important factors are temperature, pH, adsorption and sedimentation, nutrients, salinity, and biological factors (e.g., predation, competition). The modifying effect of temperature is expressed by the following equation:

$$k_T = K_{20} \times 1.047^{T-20}$$

where

K_{20} = decay rate at 20°C
T = mean water temperature in pond (°C)

The modifying effect of salinity is expressed by the following equation:

$$k_T = K_{20} \times (0.006 \times \% \text{ seawater}) \times 1.047^{T-20}$$

References

Casolari, A. 1981. A model describing microbial inactivation and growth kinetics. J. Theor. Biol. 88: 1–34.

Chick, H. 1908. An investigation of the laws of disinfection. J. Hyg. 8: 92–158.

Hoff, J. C., and E. W. Akin. 1986. Microbial resistance to disinfectants: mechanisms and significance. Environ. Health Perspect. 69: 7–13.

Ray, B. T. 1995. *Environmental Engineering*, PWS, Boston.

Rubin, A. J., J. P. Engel, and D. J. Sproul, 1983. Disinfection of amoebic cysts in water with free chlorine. J. Water Pollut. Control Fed. 55: 1174–1182.

Sundstrom, D. W., and H. E. Klei. 1979. *Wastewater Treatment*, Prentice Hall, Upper Saddle River, NJ.

Weber, W. J., Jr. Ed. 1972. *Physicochemical Processes for Water Quality Control*, Wiley-Interscience, New York.

CHICK–WATSON RELATIONSHIP

Introduction

As regards disinfection of water and wastewater, we have discussed elsewhere Chick's law and the disinfectant concentration–time product *Ct*. In 1908, H. E. Watson showed that there was a relationship between the degree of disinfection shown in Chick's law and *Ct*.

Formula

$$-\ln \frac{X}{X_0} = \Lambda C t$$

where

X = concentration of living microorganisms at time t (number/unit volume)
X_0 = initial concentration of living microorganisms (number/unit volume)
Λ = coefficient of specific lethality ($L \cdot min/mg$)
C = disinfectant concentration (mg/L)
t = contact time required to kill a certain percentage of the population (min)

References

Ray, B. T. 1995. *Environmental Engineering*, PWS, Boston.
Watson, H. E. 1908. A note on the variation of the degree of disinfection with change in the concentration of the disinfectant. J. Hyg. 8: 536–539.

CHLOROPHYLL

Introduction

Algae and higher plants contain pigments such as chlorophylls (chlorophyll *a*, *b*, and *c*), xanthophylls, and carotenes. Chlorophyll *a* is a good indicator of algal biomass in aquatic environments. There are two approaches for determining chlorophyll in aquatic samples: spectrophotometric and fluorometric methods. Another approach that may be used is analysis via high-performance liquid chromatography (HPLC) (American Public Health Association, 1992; Wetzel and Likens, 1991).

Pheophytin, a degradation production of chlorophyll *a* (loss of Mg ion from the ring structure) interferes with chlorophyll *a* measurements. Therefore, chlorophyll *a* values must be corrected for pheophytin. Chlorophyll *a* degrades to pheophytin upon acidification of the sample. The chlorophyll extract is acidified with hydrochloric acid to a concentration of no more than $3 \times 10^{-3} M$ acid. Chlorophyll is extracted with 90% acetone and the absorbance of the extract is measured at 664 nm before acidification and at 665 nm after acidification.

Formulas

Determination of Chlorophyll a in the Presence of Phaeophytin The optical density of the acetone extract is read at 750 and 664 nm before acidification, and at 750 and 665 nm after acidification. Afterward, the OD_{750} value is subtracted from OD_{664} and OD_{665}, before and after acidification, respectively. Chlorophyll *a* concentration (mg/m^3) is given by

$$C\ (\mathrm{mg/m^3}) = \frac{26.7(\mathrm{OD}_{664b} - \mathrm{OD}_{665a})V_1}{V_2 L}$$

where

C = chlorophyll *a* concentration corrected for pheophytin (mg/m^3)
OD_{664b} = optical density of extract at 664 nm before acidification
OD_{665a} = optical density of extract at 665 nm after acidification
V_1 = volume of acetone extract (L)
V_2 = volume of sample (m^3)
L = light path length or width of cuvette (cm)

Determination of Chlorophyll a, b, and c by the Spectrophotometric Method (Trichromatic Method) (American Public Health Association 1992) Briefly, the water sample is passed through a membrane filter and then extracted with 90% aqueous acetone. The optical density of the extract is read at wavelengths of 664, 647, and 630 nm for determination of chlorophyll *a*, *b* and *c*, respectively. The three optical densities are corrected for turbidity by subtracting the absorbance at 750 nm.

$$C_a(\mathrm{mg/L}) = 11.85(\mathrm{OD}_{664}) - 1.54(\mathrm{OD}_{647}) - 0.08(\mathrm{OD}_{630})$$
$$C_b(\mathrm{mg/L}) = 21.03(\mathrm{OD}_{647}) - 5.43(\mathrm{OD}_{664}) - 2.66(\mathrm{OD}_{630})$$
$$C_c(\mathrm{mg/L}) = 24.52(\mathrm{OD}_{630}) - 7.60(\mathrm{OD}_{647}) - 1.67(\mathrm{OD}_{664})$$

where

C_a, C_b, C_c = concentrations of chlorophylls *a*, *b*, and *c*, respectively (mg/L)
OD_{664}, OD_{647}, OD_{630} = turbidity-corrected optical densities (1-cm light path) at wavelengths of 664, 647 and 630 nm, respectively

The chlorophyll concentration in a given water sample is given by the following (chlorophyll *a* is given as an example):

$$\text{chlorophyll } a \text{ concentration } (\mathrm{mg/m^3}) = \frac{C_a(\mathrm{mg/L}) \times \text{extract volume (L)}}{\text{volume of sample } (\mathrm{m^3})}$$

TABLE C1: Chlorophyll *a* of Lakes and Reservoirs

	Oligotrophic	Mesotrophic	Eutrophic
Chlorophyll *a* (mg/m^3)	1.7	4.7	14.3
	(0.3–4.5)	(3–11)	(3–78)
Chlorophyll *a* peaks (mg/m^3)	4.2	16.1	42.6
	(1.3–10.6)	(4.9–49.5)	(9.5–275)

Source: Data from Vollenweider (1979), Wetzel (1983).

Numerical Values

Typical chlorophyll *a* concentrations are listed in Table C1.

References

American Public Health Association. 1992. *Standard Methods for the Examination of Water and Wastewater*, 18th ed., APHA, Washington, DC.

Strickland, J. D. H., and T. R. Parsons. 1968. *A Practical Handbook of Sea Water Analysis*, Fisheries Research Board of Canada, Ottawa, Ontario, Canada.

Vollenweider, R. A. 1979. Däs Nahrstroffbelastungskonzept als Grundlage für den externen Eingriff in den Eutrophierungsprozess stehender Gewässer und Talsperren. Z. Wasser Abwasser Forsch. 12: 46–56.

Wetzel, R. G. 1983. *Limnology*, 2nd ed., Saunders College Publishing, Philadelphia.

Wetzel, R. G., and G. E. Likens. 1991. *Limnological Analyses*, 2nd ed., Springer-Verlag, New York.

CLEARANCE RATES: Zooplankton and Protozoa

Introduction

Zooplankton and protozoa graze on phytoplankton or bacterioplankton in aquatic environments. These organisms remove particles by filtration. The clearance rate (i.e., *grazing rate*) is defined as the volume of water cleared per unit time. The clearance rate is obtained by measuring the rate of removal of bacterial cells or other particles by zooplankton or protozoa. Bacterial numbers may be determined by the acridine orange direct count (AODC) method (Hobbie et al., 1977). Clearance rates can also be obtained by using radioactively labeled food (algae, bacteria, yeast) (Knoechel and Holtby, 1986).

Formula

Clearance rate CR (mL cleared/animal per hour) is he volume of water cleared of prey (bacteria or algae) per unit time (Wetzel and Likens, 1991). Clearance rate is sometimes also called grazing rate.

$$\mathrm{CR} = \frac{C_0 - C_t}{C_0} \times \frac{v}{n} \times \frac{1}{t}$$

where

C_0 = bacterial numbers at time = 0 (cells/mL)

C_t = bacterial numbers at time t (cells/mL) (generally, grazing of zooplankton on bacteria is allowed to proceed for 10 min)

v = volume of water (mL)

n = number of animals

t = time (h)

The daily clearance rate CR of a zooplankton fed radioative-labeled food is the following (Knoechel and Holtby, 1986):

$$\mathrm{CR\ (mL/animal \cdot day)} = \frac{\text{animal (cpm/animal)}}{\text{chamber (cpm/mL)}} \times \frac{\text{1440 (min/day)}}{\text{exposure time (min)}}$$

Numerical Values

Zooplankton

- *Calamoecia lucasi*: 2.16 mL/animal · day (Forsyth and James, 1984).
- *Ceriodaphnia dubia*: 8.4 mL/animal · day (Forsyth and James, 1984).
- *Ceriodaphnia lacustris* (feeding on natural bacterioplankton): 2.4–7.2 mL/animal · day (Porter, 1084).
- *Daphnia* spp.: 18.0–33.4 mL/animal · day (Peterson et al., 1978).
- *Daphnia magna*: 66.4 mL/animal · day (Porter et al., 1983).
- *Penilia avirostris* (marine cladoceran) (Wong et al., 1992): June: CR = 2.2 mL/animal · day (0.2–20.3) (28 stations investigated); November: CR = 0.8 mL/animal · day (range: 0.1–3.4) (28 stations investigated).
- *Bosmina longirostris* (feeding on natural bacterioplankton): 5.04 mL/animal · day (Porter, 1984).

- *Copepod* clearance rates range from 0.6 to 4.8 mL/animal · day (data summarized by Cohen et al., 1984); copepods feeding on protozoa (*Monas* s.p.): 1.9–9.8 mL/animal · day (Stoecker and Capuzzo, 1990).
- *Rotifers* (*in situ* clearance rate; feeding on labeled bacteria, yeast, or algae): clearance rate is less than 10 µL/animal · h (range between 0 and 8 µL/animal · h) (Bogden et al., 1980). For rotifers feeding on fluorescent microspheres, the clearance rates varied between 3.8 nL/individual · h and 581 nL/individual · h (Sanders et al., 1989).

Protozoa

- 1–2 $\times 10^{-5}$ mL seawater/h per zooflagellate feeding on bacteria in seawater samples from Aarhus Bay, Denmark (Anderson and Fenchel, 1985).
- Heterotrophic nanoflagellates from Lake Arlington, Texas: 0.5–25 nL/nanoflagellate · h (this is equivalent to up to 58 bacterial/nanoflagellate · h) (Chrzanwoski and Simek, 1993). For protozoan flagellates feeding on fluorescent microspheres, the clearance rates varied between 0.2 and 44.4 nL/individual · h (Sanders et al., 1989).
- Laboratory feeding studies with *Monas* sp. feeding on bacteria (*Escherichia coli*, *Salmonella typhimurium*, *Chlorobium phaeobacterioides*, and an unidentified bacterial isolate from Lake Kinneret, Israel): The grazing rate varied from $<2.0 \times 10^{-7}$ to 9.5×10^{-7} mL/flagellate · h (Sherr et al., 1983).
- *Tetrahymena pyriformis* (feeding on natural bacterioplankton): 10^{-6}–10^{-5} mL/protozoan · h (Porter, 1984).
- For ciliates feeding on fluorescent microspheres, the clearance rates varied between 8.8 and 222 nL/individual · h (Sanders et al., 1989).

Clearance rate (CR) is related to the ingestion rate (I) by the following formula (Wong et al., 1992):

$$\mathrm{CR} = \frac{1}{F}$$

where

I = ingestion rate (cells/animal · h)
F = food concentration (cells/mL)
CR = clearance rate (mL water cleared/animal · h)

References

Anderson, P., and T. Fenchel. 1985. Bacterivory by microheterotrophic flagellates in seawater samples. Limnol. Oceanogr. 30: 198–202.

Bogdan, K. G., J. J. Gilbert, and P. L. Starkweather. 1980. *In situ* clearance rates of planktonic rotifers. Hydrobiologia 73: 73–77.

Chrzanowski, T. H., and K. Simek. 1993. Bacterial growth and losses due to bacterivory in a mesotrophic lake. J. Plankton Res. 15: 771–785.

Cohen, R. R. H., P. V. Dresler, E. J. P. Phillips, and R. L. Cory. 1984. The effect of the asiatic clam, *Corbicula fluminea*, on phytoplankton of the Potomac River, Maryland. Limnol. Oceanogr. 29: 170–180.

Forsyth, R. J., and M. R. James. 1984. Zooplankton grazing on lake bacterioplankton and phytoplankton. J. Plankton Res. 6: 803–810.

Hobbie, J. E., R. J. Daly, and S. Jasper. 1977. Use of Nuclepore filters for counting bacteria by fluorescence microscopy. Appl. Environ. Microbiol. 33: 1225–1228.

Jones, J. G., and G. M. Simon. 1975. An investigation of errors in direct counts of aquatic bacteria by epifluorescence microscopy, with reference to a new method for dyeing membrane filters. J. Appl. Bacteriol. 39: 317–329.

Knoechel, R., and L. B. Holtby. 1986. Cladoceran filtering rate: body length relationships for bacterial and large algal particles. Limnol. Oceanogr. 31: 195–200.

Peterson, B. J., J. E. Hobbie, and J. F. Haney. 1978. *Daphnia* grazing on natural bacteria. Limnol. Oceanogr. 23: 1039–1044.

Porter, K. G. 1984. Natural bacteria as food resources for zooplankton, pp. 340–344. In: *Current Perspectives in Microbial Ecology*, M. J. Klug and C. A. Reddy, Eds., American Society for Microbiology, Washington, DC.

Porter, K. G., Y. S., Feig, and E. F. Vetter. 1983. Morphology flow regime and filtering rates of *Daphnia, Ceriodaphnia* and *Bosmina* fed natural bacteria. Oecologia 58: 156–163.

Sanders, R. W., K. G. Porter, S. Bennett, and A. E. DeBiase. 1989. Seasonal patterns on bacterivory by flagellates, ciliates, rotifers and cladocerans in a freshwater planktonic community. Limnol. Oceanogr. 34: 673–687.

Sherr, B. F., E. B. Sherr, and T. Berman. 1983. Grazing, growth, and ammonium excretion rates of heterotrophic microflatellate fed with four species of bacteria. Appl. Environ. Microbiol. 45: 1196–1201.

Stoecker, D. K., and J. M. Capuzzo. 1990. Predation on protozoa: its importance to zooplankton. J. Plankton Res. 12: 891–901.

Wetzel, R. G., and G. E. Likens. 1991. *Limnological Analyses*, 2nd ed., Springer-Verlag, New York.

Wong, C. K., A. L. Chan, and K. W. Tang. 1992. Natural ingestion rates and grazing impact of the marine cladoceran *Penilia avirostris* Dana in Tolo harbour, Hong Kong. J. Plankton Res. 14: 1757–1765.

C/N RATIO OF MICROORGANISMS

Numerical Values

Bacterioplankton The C/N ratios of bacterioplankton from freshwater and marine environments, as determined from biovolume and biomass measurements of bacterial cells, are given in Table C2.

Phytoplankton (Stoecker and Capuzzo, 1990)

- Recommended average: C/N = 6:1
- Range: C/N = 3:1–15:1

C/N Ratios of Some Algae (Burns and Xu, 1990)

- *Anabaena flos-aquae*: 3.94
- *Nostoc Calcicola*: 4.51
- *Cyclotella* sp.: 6.03
- *Cryptomonas ovata palustris*: 8.54

Protozoan Flagellates

- *Paraphysomonas imperforata*: 4.6:1 to 7.6:1 (Goldman et al., 1985)
- *Monas* sp.: 4.6:1 (Borsheim and Bratbak, 1987)

TABLE C2: C/N Ratios of Bacterioplankton

Source	C/N Ratio (Range)	Reference
Pure culture/seawater	5.5 (4.6–6.7)	Bratbak (1985)
Fresh water	4.8 (3.3–6.8)	Nagata (1986)
Pure culture	3.5 (2.8–5.5)	Kogure and Koike (1987)
Seawater	6.3 (3.4–12.2)	Kogure and Koike (1987)
Seawater	3.7 (2.5–4.3)	Lee and Fuhrman (1987)
Fresh water	5.5 (3.4–10.5)	Nagata and Watanabe (1990)

Source: Data summarized by Lee (1993).

References

Borsheim, K. Y., and G. Bratbak. 1987. Cell volume to cell carbon conversion factors for a bacterivorous *Monas* sp. enriched from seawater. Mar. Ecol. Prog. Ser. 36: 171–175.

Bratbak, G. 1985. Bacterial biovolume and biomass estimation. Appl. Environ. Microbiol. 49: 1488–1493.

Burns, C. W., and Z. Xu. 1990. Calanoid copepods feeding on algae and filamentous cyanobacteria: rate of ingestion, defaecation and effects on trichome length. J. Plankton Res. 12: 201–213.

Goldman, J. C., D. A. Caron, O. K. Andersen, and M. R. Bennett. 1985. Nutrient cycling in a microflagellate food chain. I. Nitrogen dynamics. Mar. Ecol. Prog. Ser. 24: 231–242.

Kogure, K., and I. Koike. 1987. Particle counter determination of bacterial biomass in seawater. Appl. Environ. Microbiol. 53: 274.

Lee, S. 1993. Measurement of carbon and nitrogen biomass and biovolume from naturally derived marine bacterioplankton, pp. 319–325. In: *Handbook of Methods in Aquatic Microbiol Ecology*, P. F. Kem, B. F. Sherr, E. B. Sherr, and J. J. Cole, Eds., Lewis Publishers, Boca Raton, FL.

Lee, S., and J. A. Fuhrman. 1987. Relationships between biovolume and biomass of naturally derived bacterioplankton. Appl. Environ. Microbiol. 53: 1298–1305.

Nagata, T. 1986. Carbon and nitrogen content of natural planktonic bacteria. Appl. Environ. Microbiol. 52: 28–32.

Nagata, T., and Y. Watanabe. 1990. Carbon- and nitrogen-to-volume ratios of bacterioplankton grown under different nutritional conditions. Appl. Environ. Microbiol. 56: 1303–1309.

Stoecker, D. K., and J. M. Capuzzo. 1990. Predation on protozoa: its importance to zooplankton. J. Plankton Res. 12: 891–901.

COEFFICIENT OF CROWDING

See DEEVEY'S COEFFICIENT OF CROWDING.

COLORIFORM DIE-OFF KINETICS

Introduction

Several kinetic models for predicting the die-off of indicator bacteria in waste stabilization ponds have been proposed. First-order kinetics (*see* CHICK'S LAW) have been used to predict bacterial die-off in oxidation ponds. There are, however, several factors that affect bacterial decay rates. These factors include mainly temperature, solar radiation, predation, and antibiosis resulting from algal growth in the pond. Several investigators have proposed models based mainly on the effects of temperature and solar radiation.

Formula

Some models show a first-order equation where the decay rate is temperature dependent. Fecal coliform reduction in a completely mixed pond is given by the following equation (Marais, 1974):

$$N_e = \frac{N_i}{kR + 1} \tag{1}$$

where

N_e = coliform numbers in the effluent (numbers/mL)
N_i = coliform numbers in influent wastewater (number/mL)
k = decay rate (day^{-1})
R = retention time based on influent flow (days)

There is a relationship between the decay rate k and temperature:

$$k_T = k_{20}\theta^{T-20} \tag{2}$$

where

k_{20} = decay rate at 20°C (day^{-1})
θ = constant
T = mean water temperature in pond (°C)

In a completely mixed pond within the temperature range 5–21°C, k_T is given by the following equation:

$$k_T = 2.6(1.19)^{T-20} \tag{3}$$

In a series of n ponds, equation (1) becomes

$$N_e = \frac{N_i}{(kR_1 + 1)(kR_2 + 1)\cdots(kR_n + 1)} \tag{4}$$

where

R_1, R_2, R_n = retention times in pond $1, 2, \ldots, n$, respectively
n = number of ponds in the series (it is assumed that all ponds have the same size)

For a fixed total retention time, the removal efficiency increases as the number of ponds in the series increases.

Solar radiation also greatly influences the rate of coliform decay in oxidation ponds (Sarikaya and Saarci, 1987). The relationship between the decay rate k and light intensity I is given by

$$k = k_d + k_s(I) \tag{5}$$

where

k_d = decay rate in the dark for $I = 0$ (day^{-1}); k_d is temperature dependent
k_s = decay rate due to the effect of light (day^{-1})
I = light intensity ($\text{cal/cm}^2 \cdot \text{day}$)

Sarikaya and Saarci (1987) showed that the relationship between k and I at temperatures ranging from 25 to 30°C is given by

$$k = 0.018 + 0.012(I) \tag{6}$$

Pond depth also has a significant effect on bacterial decay in waste stabilization ponds. The effect of pond depth on coliform decay rate constant is given by (Sarikaya et al., 1987)

$$k = 1.156 + 5.244 \times 10^{-3} \frac{S_0}{KH}(1 - e^{-KH}) \tag{7}$$

where

k = decay rate (day^{-1})
S_0 = daily solar radiation ($\text{cal/cm}^2 \cdot \text{day}$)
K = light attenuation coefficient (m^{-1})
H = pond depth (m)

Equation (7) shows that bacterial decay rates in shallow ponds are higher than in deeper ponds.

A multiple linear regression equation gives the bacterial die-off rate k as a function of temperature as well as algal concentration and influent COD loading rate (Polprasert et al., 1983):

$$k = f(T, C_s, \text{OL})$$

where

T = temperature (°C)
C_s = algal concentration (mg/L)
OL = COD loading rate (kg COD/ha · day)

k is given by the following equation (Polprasert et al., 1983):

$$e^k = 0.6351(1.0281)^T(1.0016)^{C_s}(0.9994)^{\mathrm{OL}}$$

References

Marais, G. R. 1874. Fecal bacterial kinetics in stabilization ponds. J. Sanit. Eng. Div. ASCE 100: 119–139.

Polprasert, C., M. G. Dissanayake, and N. C. Thanh. 1983. Bacterial die-off kinetics in waste stabilization ponds. J. Water Pollut. Control Fed. 55: 285–296.

Sarikaya, H. Z., and A. M. Saarci. 1987. Bacterial die-off in waste stabilization ponds. J. Environ. Eng. Div. ASCE 113: 366–382.

Sarikaya, H. Z., A. M. Saarci, and A. F. Abdulfattah. 1987. Effect of pond depth on bacterial die-off. J. Environ. Eng. Div. ASCE 113: 1350–1362.

CONCENTRATION OF CONTAMINANT IN RECEIVING WATERS

Introduction

The discharge of wastewater into a receiving water increases the concentration of any pollutant above the ambient concentration. The increment of pollutant concentration above background is decreased by the dilution factor S, or is increased by the relative wastewater concentration p.

Formulas

$$C = C_s + \frac{1}{S}(C_D - C_s)$$

where

C_s = background concentration of substance X in ambient water
C_D = concentration of X in the wastewater discharge
S = dilution factor

Also:

$$C = C_s + p(C_D - C_s)$$

where p is the relative wastewater concentration.

Reference

Fischer, H. B., E. J. List, R. C. Y. Koh, J. Imberger, and N. H. Brooks. 1979. *Mixing in Inland and Coastal Waters*, Academic Press, San Diego, CA, chap. 1.

CONDUCTIVITY, HYDRAULIC

See DARCY'S LAW.

CONSUMPTION EFFICIENCY

See ECOLOGICAL EFFICIENCIES.

CONTINUOUS CULTURE OF MICROORGANISMS: Microbial Growth in an Open System

Introduction

In continuous culture of microorganisms, the nutrients are continuously added to the reactor. Assuming no biomass input, the biomass balance in the culture vessel is shown as follows (Figure C3):

$$\begin{pmatrix}\text{rate of change of}\\ \text{biomass concentration}\\ \text{in the culture vessel}\end{pmatrix} = \begin{bmatrix}\text{rate of biomass}\\ \text{production (growth)}\end{bmatrix} - \begin{bmatrix}\text{rate of biomass}\\ \text{removal (washout)}\end{bmatrix}$$

Formulas

$$\begin{aligned}\frac{dX}{dt} &= \mu X - DX\\ &= (\mu - D)X\end{aligned}$$

Assuming Monod growth kinetics (*see* MONOD'S EQUATION), the equation above gives

$$\frac{dX}{dt} = \frac{\mu_{\max} SX}{K_s + S} - DX$$

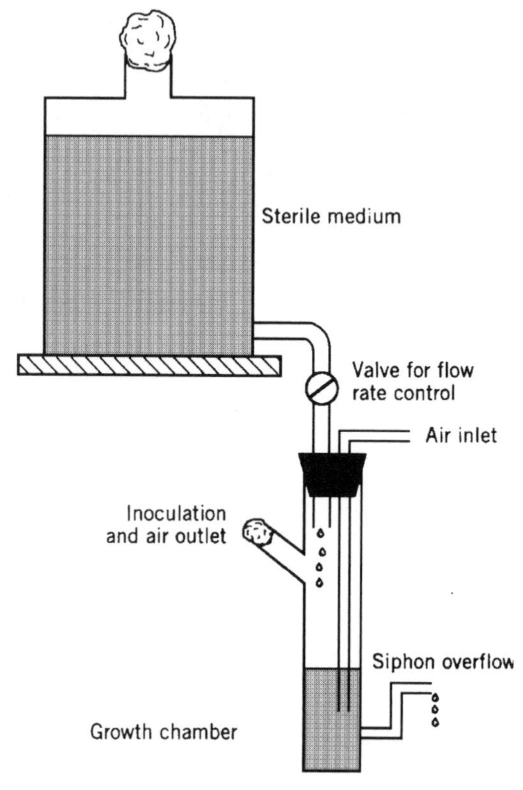

FIGURE C3: Chemostat for Continuous Culture of Microorganisms (*Source:* Adapted from Marison, 1988.)

where

dX/dt = rate of change in cell biomass (g/L · h)
μ = specific growth rate (h^{-1})
S = concentration of growth limiting substrate (mg/L)
X = cell biomass in the culture vessel (g)
D = dilution rate (h^{-1})
K_s = saturation constant (mg/L)

There are three possible situations:

1. $\mu > D \rightarrow dX/dt$ positive indicates an increase in biomass concentration.
2. $\mu < D \rightarrow dX/dt$ negative indicates a decrease in biomass concentration.
3. $\mu = D \rightarrow dX/dt = 0$ indicates a steady-state situation (this situation is preferred).

The dilution rate D is given by the following formula:

$$D = \frac{F}{V}$$

where

D = dilution rate (h^{-1})
F = flow rate of growth medium (mL/h)
V = volume of the reactor (mL)

For example, $D = 1\,h^{-1}$ means that there is one complete volume change per hour.

In a chemostat, steady state is reached by controlling the availability of the growth-limiting substrate. At steady state:

$$D(S - S_0) = \frac{\mu X}{Y} = \frac{\mu_{max} SX}{Y(K_s + S)}$$

where

D = dilution rate (h^{-1})
S = substrate concentration in the culture vessel (mg/L)
S_0 = substrate concentration in the feed solution (mg/L)
μ = specific growth rate (h^{-1})
Y = yield coefficient

Since at steady state $\mu = D \rightarrow X = Y(S_0 - S)$.

References

Bazin, M., and A. Menell. 1990. Mathematical methods in microbial ecology. Methods Microbiol. 22: 125–179 (R. Grigorova and J. R. Norris, Eds.), Academic Press, London.

Drew, S. W. 1981. Liquid culture, pp. 151–178. In: *Manual of Methods for General Bacteriology*, P. Gerhardt et al., Eds., American Society for Microbiology, Washington, DC.

Marison, L. W. 1988. In: *Biotechnology for Engineers: Biological Systems in Technological Processes*, A. Scragg, Ed., Ellis Horwood, Chichester, West Sussex, England.

Slater, J. H. 1979. Microbial population and community dynamics, pp. 45–63. In: *Microbial Ecology: A Conceptual Approach*, J. M. Lynch and N. J. Poole, Eds., Blackwell Scientific, Oxford.

D

DALTON'S LAW

Definition

Ideal mixtures of different gases in a confined volume will exert individual pressures directly proportional to their molar fractions in that volume. The total pressure will be the sum of those pressures.

Formula

$$P_T = \sum_{i=1}^{n} P_i$$

where

P_T = total pressure of the gas (atm)

P_i = partial pressure of any gas i in the volume (atm)

Reference

Wanielista, M. P., Y. A. Yousef, J. S. Taylor, and C. D. Cooper. 1984. *Engineering and the Environment*, Brooks/Cole, Pacific Grove, CA.

DARCY'S LAW

Introduction

Darcy's law is an empirical law that describes the flow of water through porous media (e.g., aquifers, soils). It was formulated in Dijon, France, in 1856, by Henri Darcy, a French hydraulic engineer. This law is widely used in hydrology, agricultural engineering, chemical engineering, environmental engineering, soil physics, and soil mechanics (Freeze and Cherry, 1979).

Formula

Figure D1 shows a vertical cross section of groundwater flow with linear parallel streamlines. If piezometers are placed at points 1 and 2, the velocity of groundwater in a given streamline is given by Darcy's equation:

$$v = K\frac{H_1 - H_2}{L}$$

where

v = Darcy's velocity or *specific discharge* (length/time)
K = hydraulic conductivity (length/time)
$H_1 = h_1 + z_1$ h_1 = pressure head or fluid levels at point 1 (length)
z_1 = elevation head at point 1 (length)
$H_2 = h_2 + z_2$ h_2 = pressure head or fluid levels at point 2 (length)
z_2 = elevation head at point 2 (length)
L = distance between points 1 and 2 along the streamline

Thus the velocity of water v is proportional to the hydraulic gradient, $(H_1 - H_2)/L$.

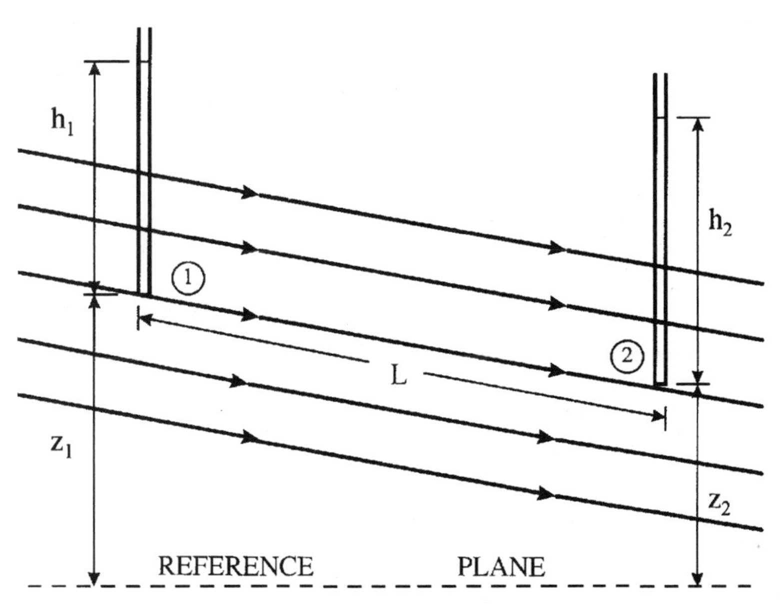

FIGURE D1: Vertical Cross Section of Groundwater Flow with Linear, Parallel Streamlines (*Source:* Bouwer, 1978.)

TABLE D1: Hydraulic Conductivity of Some Soil and Aquifer Materials

Soil or Aquifer Material	K (m/day)
Clay soils (surface)	0.01–0.2
Loamy soils (surface)	0.1–1.0
Fine sand	1–5
Medium sand	5–20
Coarse sand	20–100
Gravel	100–1000

Source: Data from Bouwer (1978), Davis (1969).

Numerical Values

The constant of proportionality K or *hydraulic conductivity* is generally high for sands and gravels and low for clays. Some values of K are listed in Table D1.

References

Bouwer, H. 1978. *Groundwater Hydrology*, McGraw-Hill, New York.

Davis, S. N. 1969. Porosity and permeability of natural materials, pp. 54–89. In: *Flow Through Porous Media*, R. J. M. de Wiest, Ed., Academic Press. San Diego, CA.

Freeze, R. A., and J. A. Cherry. 1979. *Groundwater*, Prentice Hall, Upper Saddle River, NJ.

DECAY RATE OF MICROORGANISMS

See also CHICK'S LAW.

Introduction

Microorganisms can be inactivated following exposure to environmental factors (e.g., temperature, desiccation, light), sterilization, or disinfection.

Formula

$$N_t = N_0 e^{-kt}$$

where

N_t = number of microorganisms after time t (number/mL)

N_0 = initial number of microorganisms (number/mL)

k = decay rate constant (min^{-1})

t = exposure time (min)

$-k$ is the slope of a semilog plot of log number of microorganisms versus time. k can be used to derive the *decimal reduction factor* D, which is the time to reduce a viable microbial population by 90% or one order of magnitude. D is used for the determination of the appropriate temperature and exposure time for a food preservation process in the food industry.

$$D = \frac{2.303}{k}$$

If one plots $\log_{10} N_t$ versus time, one obtains a straight line where the slope is equal to $-1/D$ and the y-intercept is equal to $\log_{10} N_0$. Figure D2 gives an example of the determination of D value concerning the survival of a spore-forming bacterium at 121°C.

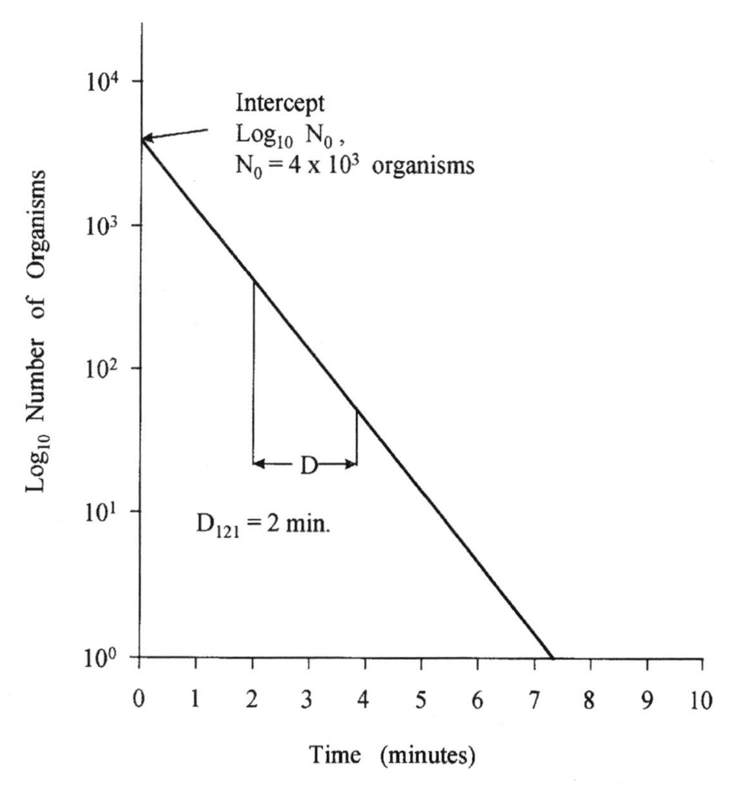

FIGURE D2: Determination of the *D* Value Concerning the Survival of a Spore Former at 121°C (*Source:* Bader, 1986.)

TABLE D2: Decay Rates of Microorganisms in Some Environments

Microorganism	Decay Rate (h^{-1})	Reference
Poliovirus 1		
Groundwater	0.0019	Bitton et al. (1983)
River water (Rio Grande)	0.031	O'Brien and Newman (1977)
Seawater	0.02	Matossan and Garabedian (1967)
Salmonella typhimurium		
Groundwater	0.0054–0.0091	Bitton et al. (1983) McFeters et al. (1974)
Escherichia coli		
Groundwater	0.0066	Bitton et al. (1983)
Groundwater (fecal coliforms)	0.0152	McFeters et al., (1974)

Numerical Values

See Table D2.

References

Bader, F. G. 1986. Sterilization: prevention of contamination, pp. 345–362. In: *Manual of Industrial Microbiology and Biotechnology*, A. L. Demain and N. A. Solomon, Eds., American Society for Microbiology, Washington, DC.

Bailey, J. E., and D. F. Ollis. 1977. *Biochemical Engineering Fundamentals*, McGraw-Hill, New York.

Bitton, G., S. R. Farrah, R. H. Ruskin, J. Butner, and Y. J. Chou. 1983. Survival of pathogenic and indicator organisms in groundwater. Ground Water 21: 405–410.

Matossan, A. M., and G. A. Garabedian. 1967. Virucidal action of seawater. Am. J. Epidemiol. 85: 1–8.

McFeters, G. A., G. K. Bissonnette, J. J. Jeseski, C. A. Thomson, and D. G. Stuart. 1974. Comparative survival of indicator bacteria and enteric pathogens in well water. Appl. Microbiol. 27: 823–829.

O'Brien, R. T., and J. S. Newman. 1977. Inactivation of polioviruses and coxsackieviruses in surface waters. Water Res. 10: 751–755.

DECIMAL REDUCTION TIME: *D* Value

see DECAY RATE OF MICROORGANISMS.

DEEVEY'S COEFFICIENT OF CROWDING

Introduction

Deevey's coefficient was derived for barnacle growth and expresses how barnacles impinge on one another as they grow (Deevey, 1947; Southwood, 1978).

Formula

$$C_c = 2\pi r^2 N^2$$

where

C_c = coefficient of crowding (number of contacts/cm^2)
r = radius of fully grown animal (cm)
N = density of animals per unit area (number/cm^2)

References

Deevey, E. S. 1947. Life tables for natural populations of animals. Q. Rev. Biol. 22: 283–314.

Southwood, T. R. E. 1978. *Ecological Methods*, Chapman & Hall, London.

DENSITY: Vertical Distribution of Densities of Water

Introduction

The vertical distribution of densities of water in a receiving water has a large influence on vertical mixing. Since density differences tend to be small (typically less than 3%), they are typically expressed in terms of the parameter s_t. Information on water density as a function of salinity and temperature is typically tabulated in terms of s_t values, particularly in oceanography literature.

Formula

The density of water (r) at 1 atm pressure is given by

$$r = 1 + \frac{s_t}{1000}$$

if r is expressed in g/cm^3, or

$$r = 1 + s_t$$

if r is expressed in kg/m^3. For example, if a water has an s_t value of 26.35, its density is $1.02635\,g/cm^3$. The density of water increases with depth due to increasing hydrostatic pressure.

Reference

Fischer, H. B., E. J. List, R. C. Y. Koh, J. Imberger, and N. H. Brooks. 1979. *Mixing in Inland and Coastal Waters*, Academic Press, San Diego, CA, chap. 1.

DENSITY (BUOYANT) OF MICROORGANISMS

Definition

Density is the mass m of microorganisms per unit volume V of microorganisms.

Formula

$$r = \frac{m}{V}$$

where

m = mass (g)
V = unit volume (cm^3)

The extracellular material of cells influences the determination of cell density.

Numerical Values

Bacteria

- *B. subtilis*: $r = 1.15$–1.24 (calculated); $r = 1.13$ (measured) (Bratbak and Dundas, 1984)
- *E. coli*: $r = 1.09$–1.13 (calculated); $r = 1.09$ (measured) (Bratbak and Dundas, 1984)
- *P. putida*: $r = 1.14$–1.22 (calculated); $r = 1.12$ (measured) (Bratbak and Dundas, 1984)
- Soil bacterial isolates: $r = 1.035$–1.093 (Bakken and Olsen, 1983)

Fungi

- Soil fungal isolates (hyphae): $r = 1.027$–1.13 (Bakken and Olsen, 1983)

References

Bakken, L. R., and R. A. Olsen. 1983. Buoyant densities and dry matter contents of microorganisms: conversion of a measured biovolume into biomass. Appl. Environ. Microbiol. 45: 1188–1195.

Bratbak, G., and I. Dundas. 1984. Bacterial dry matter and biomass estimations. Appl. Environ. Microbiol. 48: 755–757.

DENSITY INDEX

See SLUDGE DENSITY INDEX.

DIFFUSION COEFFICIENT OF BIOCOLLOIDS

See STOKES–EINSTEIN EQUATION.

DILUTED SLUDGE VOLUME INDEX (DSVI)

Introduction

The DSVI is the sludge volume index computed from a 30-minute settling test in which the mixed liquor from an activated sludge aeration tank occupies no more than 20% of the original volume of the test cylinder. The mixed liquor is diluted as necessary to achieve this result. This procedure makes the computed index independent of the original mixed liquor suspended solids concentration. DSVI was found to be the best index of sludge settleability (Lee et al., 1983).

Formula

$$\mathrm{DSVI} = \frac{\mathrm{CV}_{30}}{\mathrm{SS}_i} 2^n$$

where

DSVI = diluted sludge volume index (mL/g)

CV_{30} = compact volume of sludge after 30 minutes of settling (mL)

SS_i = initial suspended solids concentration of the mixed liquor (g/L)

n = number of twofold dilutions required to achieve a CV_{30} of 200 mL or less

Reference

Lee, S.-E., B. Koopman, H. Bode, and D. Jenkins. 1983. Evaluation of alternative sludge settleability indices. Water Res. 17: 1421–1426.

DISINFECTION: Concentration–Time Product *Ct*

Introduction

Watson's law deals with the relationship between disinfectant concentration and contact time. Disinfectant effectiveness may be expressed as Ct, C being the disinfectant concentration and t the time required to inactivate a given percentage of the population under specific conditions (pH and temperature).

Formula

$$K = C^n t$$

where

K = constant for a given microorganism exposed to a disinfectant under specific conditions (mg/L · min).

C = disinfectant concentration (mg/L)

n = empirical constant, also called the coefficient of dilution

t = contact time required to kill a certain percentage of the population (min)

When t is plotted against C on double logarithmic paper, n is the slope of the straight line (Figure D3). The value of n determines the importance of the disinfectant concentration or contact time in microorganism inactivation. However, in engineering practice, n value is often assumed to be close to unity.

References

Baumann, E. R., and D. D. Ludwig. 1962. Free available chlorine residuals for small nonpublic water supplies. J. Am. Water Works Assoc. 54: 1379–1388.

Clark, R. M., E. J. Read and J. C. Hoff. 1989. Analysis of inactivation of *Giardia lamblia* by chlorine. J. Environ. Eng. Div. ASCE 115: 80–90.

Rubin, A. J., J. P. Engel, and O. J. Sproul. 1983. Disinfection of amoebic cysts in water with free chlorine. J. Water Pollut. Control. Fed. 55: 1174–1182.

DISINFECTION: Inactivation Rate Microorganisms

See CHICK'S LAW.

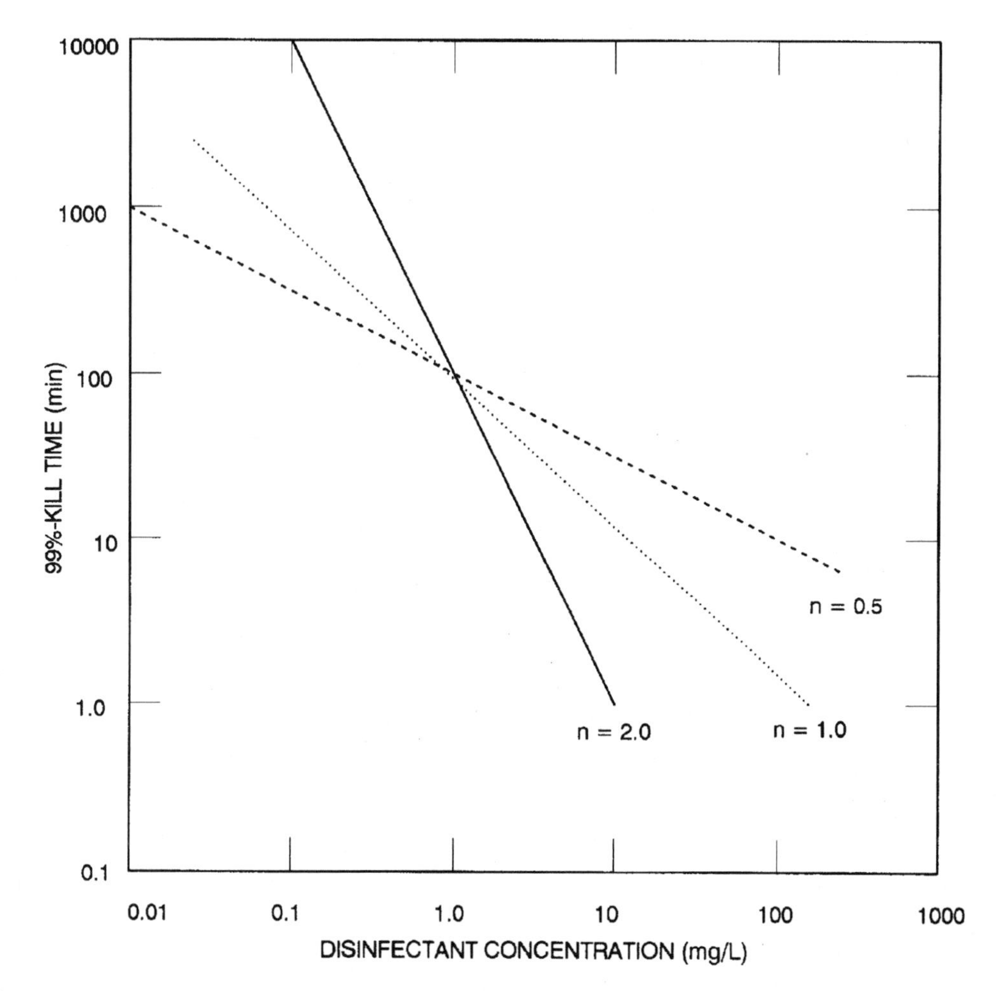

FIGURE D3: Effect of *n* value on *Ct* at Various Disinfectant Concentrations (*Source:* Adapted from Clark, et al., 1989.) Analysis of inactivation of *Giardia lamblia* by chlorine. J. Environ. Eng. Div. 115: 80–90).

DISINFECTION: Lethality Coefficient

Definition

The lethality coefficient expresses the relative efficiency of a disinfectant.

Formula

$$\lambda = \frac{4.6}{Ct_{99}}$$

TABLE D3: Values of Lethality Coefficient λ and Residual O_3 to Destroy 99% of Microorganisms in 10 Minutes at 10–15°C

Microorganisms	Value of λ	Residual O_3 (mg/L)
Escherichia coli	500	0.001
Streptococcus fecalis	300	0.0015
Poliovirus	50	0.01
Bacillus megaterium spores	15	0.03
Entamoeba histolytica cysts	5	0.1

Source: Adapted from Chang (1982).

where

4.6 = natural log of 100

C = residual disinfectant concentration (mg/L)

t_{99} = contact time to achieve 99% destruction of microorganisms (min)

Numerical Values

The value of λ for destruction of various microorganisms by ozone varies with the type of microorganism (Table D3). λ also varies with the type of disinfectant, as shown in Table D4.

References

Chang, S.-L. 1982. The safety of water disinfection. Annu. Rev. Public Health 3: 393–418.

Morris, J. C. 1975. Aspects of the quantitative assessment of germicidal efficiency. In: *Disinfection of Water and Wastewater*, J. D. Johnson, Ed., Ann Arbor Science, Ann Arbor, MI.

TABLE D4: Values of λ for 99% Destruction of Four Groups of Organisms in 1 Minute at 50°C by Ozone and Three Chlorine Compounds

Disinfection Agent	Value of λ			
	Escherichia coli	Enterovirus	Amoebic Cysts	Bacterial Spores
O_3	50	5	0.5	1.5
HOCI	20	1	0.05	0.05
OCI	0.2	0.02	0.0005	0.0005
NH_2Cl	0.1	0.005	0.02	0.001

Source: Adapted from Chang (1982).

DISSOLVED OXYGEN SAG EQUATION

See STREETER–PHELPS EQUATION.

DIVERSITY INDEX

Introduction

Stability of an ecosystem is often evaluated by measuring the number of species in a community and the number of individuals in each species. Thus species diversity takes into account the *species richness* (number of species) and *evenness* (the relative number of individuals in the species). The equitable distribution of individuals among the species leads to an increase in species diversity. The complexity of an ecosystem is expressed as a *diversity index*.

Some common diversity indexes are the Shannon–Wiener index (or Shannon–Weaver index or Shannon index), species richness, Simpson index, and Margalef index (Brillouin, 1960; Peet, 1974; Pielou, 1966, 1975). These indexes are used to determine stresses on ecosystems.

Formulas

Shannon Index, H This is also called the *Shannon–Wiener index* or *Shannon–Weaver index* (Shannon and Weaver, 1949). It is the most widely used index for measuring biological diversity:

$$H = \sum_{i=1}^{S} P_i \ln P_i$$

where

S = total number of species

$P_i = N_i/N$ = proportion of individuals of the total sample belonging to ith species

N_i = number of individuals (N) belonging to the ith species

N = total number of individuals

From the formula above we obtain the *Shannon equitability* (or evenness) *index J*:

$$J = \frac{H}{\ln S}$$

J ranges from 0 to 1.0.

Species Richness Species richness is the actual number of species present in a community (Atlas 1984; Patrick, 1949).

$$D = S$$

where

$D =$ species richness

$S =$ number of species in the community

Simpson Index (Simpson, 1949) This index measures the probability that two specimens picked at random in a community belong to different species (Krebs, 1978).

$$D = 1 - \sum_{i=1}^{s} P_i^2$$

where

$D =$ Simpson index

$S =$ number of species

$P_i =$ proportion of individuals in the community belonging to ith species

The Simpson index ranges between a value of 0 (low diversity) and a maximum of $1 - 1/S$.

Margalef Index, D (American Public Health Association, 1989; Margalef, 1968)

$$D = \frac{S - 1}{\ln N}$$

where

$D =$ Margalef index

$S =$ number of species in sample

$N =$ total number of individuals in a sample

McIntosh Index The McIntosh index is given by the following formula (McIntosh, 1967):

$$D = \sqrt{\sum_{i=1}^{S} (n_i)^2}$$

where

$D =$ McIntosh index

$S =$ number of species

$n_i =$ number of individuals in ith species

References

Abel, P. D. 1988. *Water Pollution Biology*, Ellis Horwood, Chichester, West Sussex, England.

American Public Health Association. 1989. *Standard Methods for the Examination of Water and Wastewater*, 17th ed., APHA, Washington, DC.

Atlas, R. M. 1984. Diversity of microbial communities. Adv. Microb. Ecol. 7: 1–47.

Brillouin, L. 1960. *Science and Information Theory*, 2nd ed., Academic Press, San Diego, CA.

Krebs, C. J. 1978. *Ecology: The Experimental Analysis of Distribution and Abundance*, 2nd ed., Harper & Row, New York.

Margalef, R. 1968. *Perspectives in Ecological Theory*, University of Chicago Press, Chicago.

McIntosh, P. R. 1967. An index of diversity and the relation of certain concepts to diversity. Ecology 48: 392–404.

Patrick, R. 1949. A proposed biological measure of stream conditions, based on a survey of the Conestoga Basin, Lancaster County, Pennsylvania. Proc. Acad. Nat. Sci. Phila. 101: 277–341.

Peet, R. K. 1974. The measurement of species diversity. Annu. Rev. Ecol. Syst. 5: 285–307.

Pielou, 1966. The measurement of diversity in different types of biological collections. J. Theor. Biol. 13: 131–144.

Pielou, E. C. 1975. *Ecological Diversity*, Wiley, New York.

Shannon, C. E., and W. Weaver. 1949. *The Mathematical Theory of Communication*, University of Illinois Press, Urbana, IL.

Simpson, E. H. 1949. Measurement of diversity. Nature 163: 688.

Wetzel, R. G. 1983. *Limnology*, 2nd ed., Saunders College Publishing, Philadelphia.

DIXON PLOTS

See also ENZYME INHIBITION.

Definition

The Dixon plot is a graphical method for determining the inhibitor constant K_i as well as the type of inhibition for an enzymatic reaction. In Dixon plots, $1/V$ is plotted against the inhibitor concentration [I] while the substrate concentration is constant.

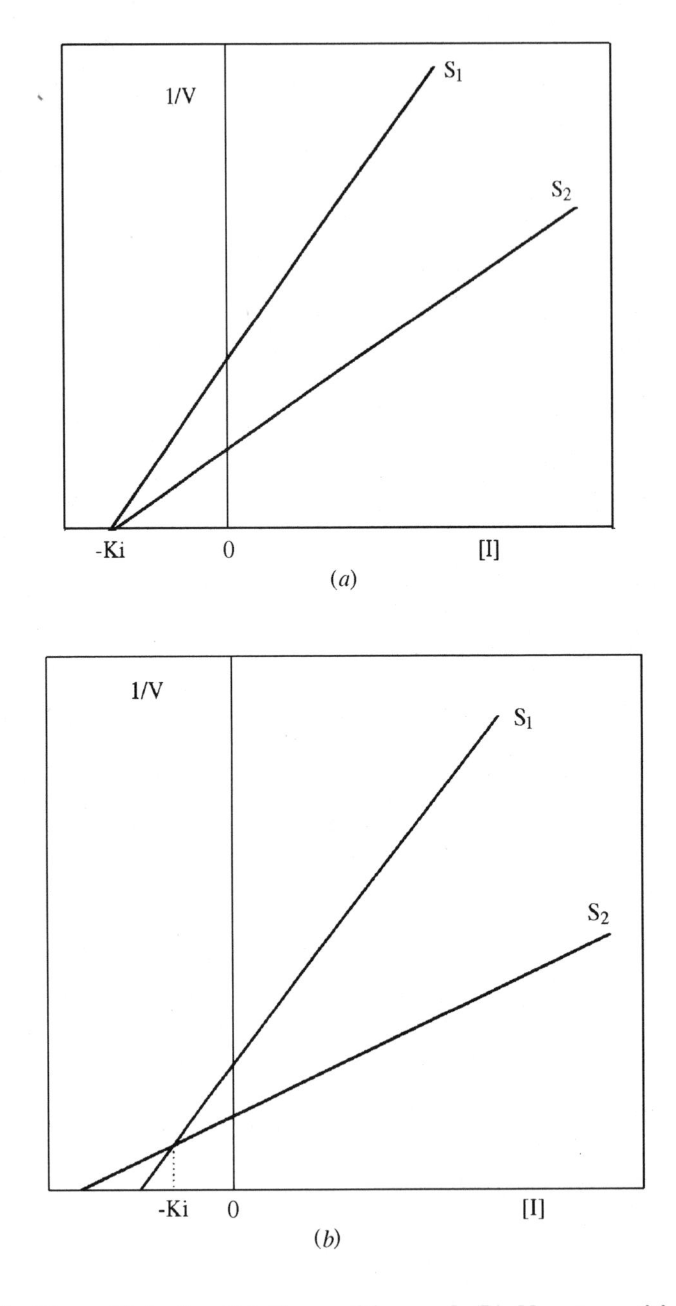

FIGURE D4: Dixon Plots for (A) Competitive and (B) Noncompetitive Enzyme Inhibition (S_1, S_2, substrate concentrations; $S_2 > S_1$) [*Source:* (B) Adapted from Segel, 1976.]

Plots

Competitive Inhibition With the substrate concentration [S] held constant, the reaction rate V is measured as a function of the inhibitor concentration [I]. $1/V$ is plotted against [I]. A straight line is obtained at each of the two or three substrate concentrations used. The straight lines intersect at an [I] value equal to $-K_i$ (Figure D4A).

Noncompetitive Inhibition Dixon plot ($1/V$*ersus* [I]) for a noncompetitive inhibitor is shown in Figure D4B. The lines at all values of [S] intersect on the negative abscissa at $[I] = -K_i$.

References

Dixon, M. 1953. The determination of enzyme inhibitor constants. Biochem. J. 55: 170–171.

Dixon, M., and E. C. Webb. 1964. *Enzymes*, 2nd ed., Academic Press, San Diego, CA.

Michal, G. 1978. Determination of Michaelis constants and inhibitor constants, pp. 29–40. In: *Principles of Enzymatic Analysis*, H. U. Bergmeyer, Ed., Verlag Chemie, Weinheim, Germany.

Segel, I. H. 1976. *Biochemical Calculations*, 2nd ed., Wiley, New York.

Zeffren, E., and P. L. Hall. 1973. *The Study of Enzyme Mechanisms*, Wiley, New York.

DNA CONTENT OF CELLS

Numerical Values

The DNA content of bacterial cells in marine waters was found to vary from 1.3 to 4.7 fg DNA/bacterial cell with an average of 2.4 fg DNA/cell (Fuhrman and Azam, 1982). Some other values that have been found are:

- Subsurface bacteria (Cape Cod, Massachusetts): 2–21 fg/cell (Metge et al. 1993)
- Marine bacteria: 5.7 fg/cell (Paul et al., 1985)

References

Fuhrman, J., and F. Azam. 1982. Thymidine incorporation as a measure of heterotrophic bacterioplankton production in marine surface waters: evaluation and field results. Mar. Biol. 66: 109–120.

Metge, T. W., M. H. Brooks, R. L. Smith, and R. W. Harvey. 1993. Effect of treated-sewage contamination upon bacterial energy charge, adenine nucleotides, and DNA content in a sandy acquifer on Cape Cod. Appl. Environ. Microbiol. 59: 2304–2310.

Paul, J. H., W. H. Jeffrey, and M. DeFlaun. 1985. Particulate DNA in subtropical and estuarine planktonic environments. Mar. Biol. 90: 94–101.

DOMINANCE INDEX, COMMUNITY

See also BERGER–PARKER DOMINANCE INDEX.

Definition

The community dominance index (CDI) is the percentage of abundance that is contributed by the two most abundant species within the community. Species abundance is determined as biomass, productivity, or density (Krebs, 1972; McNaughton, 1968).

Formula

$$\text{CDI} = \frac{y_1 + y_2}{y}$$

where

y_1 = abundance of the most abundant species
y_2 = abundance of the second-most abundant species
y = total abundance for all species

Abundance is inversely related to diversity.

References

Krebs, C. J. 1972. *Ecology*, Harper & Row, New York.

McNaughton, S. J. 1988. Structure and function in California grasslands. Ecology 49: 962–972.

DOUBLING TIME OF MICROORGANISMS

Definition

Doubling time, or *generation time*, is the time required for a cell population to double in numbers.

Formula

$$t_d = \frac{\ln 2}{\mu} = \frac{0.693}{\mu} = G$$

where

t_d = doubling time (hours or days)

ln 2 = natural log of 2 = 0.693

μ = specific growth rate (h^{-1} or day^{-1})

G = generation time (hours or days)

Numerical Values

Doubling Times Observed with Pure Cultures of Some Bacteria See Table D5.

Bacterial Doubling Times in Freshwater Environments The average doubling times of bacteria in a eutrophic lake in Norway were 31 h (range: 8–140 h) in the epilimnion and 220 h (range: 15–690 h) in the metalimnion (Vadstein et al., 1989). In eutrophic Lake Tystrup, Denmark, the bacterial generation times varied between 9 and 167 h (Riemann, 1983). In eutrophic German lakes, G ranged from 7.4 to 240 h (Krambeck et al., 1981). In a hypereutrophic lake in Sweden (Lake Vallentunasjon, Sweden) the doubling times for bacteria were 2.5–4.5 days at 20°C and 100–200 days at 4°C (Bell and Ahlgren, 1987).

Bacterial Doubling Times in Marine Environments Some mean generation times for bacteria observed in oceanic systems are shown in Table D6.

Doubling Times of Some Protozoa Turley (1993) reported a range of doubling times for water flagellates as $t_d = 0.12$–1.44 days. Laboratory feeding studies with *Monas* sp. feeding on bacteria (*Escherichia coli*, *Salmonella*

TABLE D5: Doubling Times of Some Bacterial Cultures

Organism	Temperature (°C)	Doubling Time (h)
Bacillus stearothermophilus	60	0.14
Excherichia coli	40	0.35
Pseudomonas putida	30	0.75
Vibrio marinus	15	1.35
Rhodobacter sphaeroides	30	2.2
Mycobacterium turberculosis	37	~6
Nitrobacter agilis	27	~12

Source: Adapted from Stanier et al. (1986).

TABLE D6: Doubling Times of Bacteria in Marine Environments

Region	Generation Time [Days ($\pm$SE)]
Gulf stream rings	2.8 (0.20)
North Atlantic	5.3 (0.42)
Indian Ocean	14.7 (1.7)
Subarctic North Pacific	27.8 (2.2)

Source: Adapted from Ducklow (1992).

typhimurium, *Chlorobium phaeobacterioides*, and an unidentified bacterial isolate from Lake Kinneret, Israel): $t_d = 3.2$–8.7 h. t_d was temperature dependent and was equal to 21.3, 5.4, 5.0, and 13.1 h at 3, 18, 23.5, and 30°C, respectively (Sherr et al., 1983).

References

Bailey, J. E. and D. F. Ollis. 1977. *Biochemical Engineering Fundamentals*, McGraw-Hill, New York.

Bell, R. T., and I. Ahlgren. 1987. Thymidine incorporation and microbial respiration in the surface sediment of a hypereutrophic lake. Limnol. Oceanogr. 32: 476–482.

Ducklow, H. W. 1992. Factors regulating bottom-up control of bacteria biomass in open ocean plankton communities. Arch. Hydrobiol. Beih. 37: 207–217.

Krambeck, C., H.-J. Krambeck, and J. Overbeck. 1981. Microcomputer-assisted biomass determination of plankton bacteria on scanning electron micrographs. Appl. Environ. Microbiol. 42: 142–149.

Riemann, B. 1983. Biomass and production of phyto- and bacterio-plankton in eutrophic Lake Tystrup, Denmark. Freshwater Biol. 13: 389–398.

Sherr, B. F., E. B. Sherr, and T. Berman. 1983. Grazing, growth, and ammonium excretion rates of a heterotrophic microflagellate fed with four species of bacteria. Appl. Environ. Microbiol. 45: 1196–1201.

Stanier, R. Y., L. L. Ingraham, M. L. Wheelis, and P. R. Painter. 1986. *The Microbial World*, Prentice Hall, Upper Saddle River, NJ.

Turley, C. M. 1993. Determination of pressure effects on flagellates isolated from surface water, pp. 91–96. In: *Handbook of Methods in Aquatic Microbial Ecology*, P. F. Kemp, B. F. Sherr, E. B. Sherr, J. Cole, Eds., Lewis Publishers, Boca Raton, FL.

Vadstein, O., B. O. Harkjerr, A. Jensen, Y. Olsen, and C. Reinertsen. 1989. Cycling of organic carbon in the photic zone of a eutrophic lake with special reference to the heterotrophic bacteria. Limnol. Oceanogr. 34: 840–855.

DRY WEIGHT: Microbial Cells

Introduction

The percent dry weight of microbial cells can be calculated from the cell biovolume, the cell buoyant density, and the dry matter content obtained after drying the sample at 105°C for 12 h (Bakken and Olsen, 1983).

Formula

$$\% \text{ Dry weight} = \frac{\text{DM} \times 100}{V\rho}$$

where

DM = dry matter content obtained after drying at 105°C for 12 h
V = biovolume determined microscopically
ρ = buoyant density of cells

Extracellular material content influences the percent dry weight of cells.

Numerical Values

Percent Dry Weight

- *Soil bacterial isolates:* 12–33% (an average 30% dry weight was proposed for converting bacterial biovolume into dry biomass). It is noted that the dry weight of extracellular material is not included (Blakken and Olsen, 1983).
- *Soil fungal isolates:* 18–25% (an average 21% dry weight was proposed for converting hyphal biovolume into dry biomass) (Bakken and Olsen, 1983).

Cell Carbon/Dry Weight Ratio Some data were summarized by Fry (1988):

- *Athrobacter globiformis:* 0.369
- *Enterobacter aerogenes:* 0.428
- *Bacillus subtilis:* 0.488
- *Escherichia coli:* 0.480
- *Pseudomonas putida:* 0.463

References

Bakken, L. R., and R. A. Olsen. 1983. Buoyant densities and dry matter contents of microorganisms: conversion of a measured biovolume into biomass. Appl. Environ. Microbiol. 45: 1188–1195.

Bratbak, G., and I. Dundas. 1984. Bacterial dry matter and biomass estimations. Appl. Environ. Microbiol. 48: 755–757.

Fry, J. C. 1988. Determination of biomass, pp. 27–34. In: *Methods in Aquatic Bacteriology*, B. Austin, Ed., Wiley, New York.

Luria, S. E. 1960. The bacterial protoplasm: composition and organization, pp. 1–34. In: *The Bacteria*, vol. 1, I. C. Gunsalus and R. Y. Stanier, Eds., Academic Press, San Diego, CA.

van Veen, J. A., and E. A. Paul. 1979. Conversion of biovolume measurements of soil organisms grown under various moisture tensions to biomass and their nutrient content. Appl. Environ. Microbiol. 37: 696–692.

Watson, S. W., T. J. Novitsky, H. L. Quinby, and F. W. Valois. 1977. Determination of bacterial number and biomass in the marine environment. Appl. Environ. Microbiol. 33: 940–946.

DRY WEIGHT: Relationship Between Dry Weight and Body Length of *Daphnia magna*

Formula

The relationship between dry weight and body length of *Daphnia magna* is given by the following formula (Porter et al., 1982):

$$W = 0.0119L^{2.3907} \qquad (r^2 = 0.9726, P < 0.01)$$

where

W = dry weight of *Daphnia* (mg)

L = body length (mm)

Reference

Porter, K. G., J. Gerritsen, and J. D. Orcutt, Jr. 1982. The effect of food concentration on swimming patterns, feeding behavior, ingestion, assimilation, and respiration by *Daphnia*. Limnol. Oceanogr. 27: 935–949.

E

EADIE–SCATCHARD PLOT

Definition

The Eadie–Scatchard equation concerning enzyme reaction kinetics is a linearized form of the Michaelis–Menten equation (*see* MICHAELIS–MENTEN EQUATION).

Formula

$$\frac{V}{[\mathrm{S}]} = \frac{1}{K_m} V + \frac{V_{\max}}{K_m}$$

where

V = enzymatic reaction rate (1/time)

[S] = substrate concentration (mol/L)

K_m = half saturation constant (Michaelis constant) (mol/L) (the substrate concentration at which V is equal to $V_{\max}/2$)

$V_{\max}$ = maximum specific reaction rate (1/time)

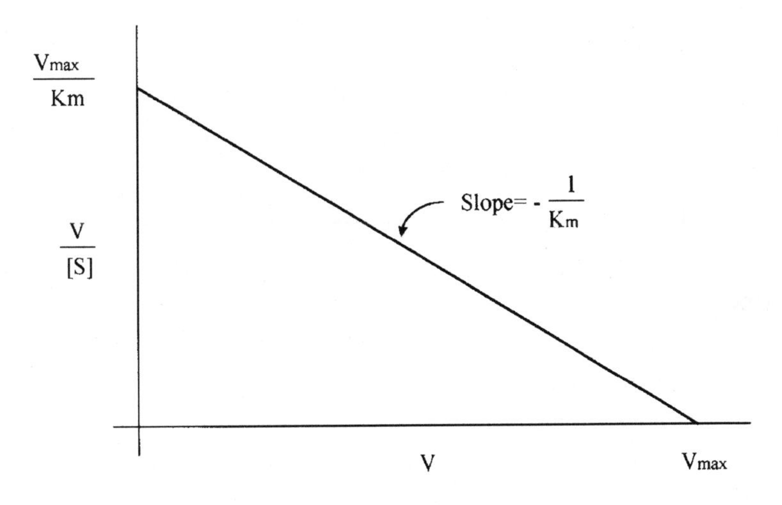

FIGURE E1: Eadie-Scatchard Plot (*Source:* Segel, 1976.)

A plot of $V/[S]$ versus V is linear and has a slope of $-1/K_m$, an intercept of V_{max}/K_m on the y-axis and an intercept of V_{max} on the x-axis (Figure E1).

References

Segel, I. H. 1976. *Biochemical Calculations*, 2nd ed., Wiley, New York.

ECOLOGICAL EFFICIENCIES

Introduction

In a community, one can measure ecological efficiencies *within* a trophic level (e.g., growth efficiency) or *between* trophic levels (e.g., consumption efficiency or Lindeman's efficiency) (Kozolovsky, 1968; Krebs, 1972).

Formula and Numerical Values

$$\text{Growth efficiency} = \frac{\text{net productivity at trophic level } n}{\text{assimilation at trophic level } n}$$

Growth efficiency is greater for plants than for animals.

$$\text{Consumption efficiency} = \frac{\text{intake at trophic level } n}{\text{net productivity at trophic level } n-1}$$

Consumption efficiency falls between 20 and 25%.

$$\text{Lindemann's efficiency} = \frac{\text{assimilation at trophic level } n}{\text{assimilation at trophic level } n-1}$$

Lindeman's efficiency is generally around 10% but can reach 70% in some cases (Petipa et al., 1970).

References

Kozlovsky, D. G. 1968. A critical review of the trophic level concept. I. Ecological efficiencies. Ecology 49: 48–60.

Krebs, C. J. 1972. *Ecology*, Harper & Row, New York.

Petipa, T. S., E. V. Pavlova, and G. N. Mironov. 1970. The food web structure, utilization and transport of energy by trophic levels in the planktonic communities, pp. 142–167. In: *Marine Food Chains*, J. H. Steele, Ed., University of California Press, Berkely, CA.

EGESTION (EVACUATION) RATE OF ZOOPLANKTON

See INGESTION RATE: Zooplankton on Protozoa.

Definition

The egestion (evacuation) rate of zooplankton is the amount of prey (bacteria or algae) egested per unit time.

Formula

Wong et al. (1992) assumed that the evacuation (egestion) of the gut algal pigment content is approximately exponential, and calculated the egestion rates (use of the marine cladoceran *Penilia avirostris*) by regression from the equation

$$G_t = G_0 e^{-Rt}$$

where

G_t = gut pigment content at time t
G_0 = initial gut pigment content
R = gut egestion rate (1/time)
t = time

If we assume that food ingestion and egestion are in equilibrium (Wong et al., 1992):

$$I = RG$$

where

I = ingestion rate (1/time)
R = egestion rate (1/time)
G = gut pigment content

Numerical Values

For *Penilia avirostris*, R varied from $0.019\,\text{min}^{-1}$ under laboratory conditions to $0.032\,\text{min}^{-1}$ under field conditions.

Reference

Wong, C. K., A. L. Chan, and K. W. Tang. 1992. Natural ingestion rates and grazing impact of the marine cladoceran *Penilia avirostris* Dana in Tolo harbour, Hong Kong. J. Plankton Res. 14: 1757–1765.

EINSTEIN ENERGY: Mass Equivalence Formula

Introduction

The mass equivalence formula is used to determine the energy of the radiation (e.g., α and β particles) of radioactive materials.

Formula

$$E = Mc^2$$

where

E = energy (g · cm/s or ergs)
M = mass of particles (g)
c = velocity of light = 2.998×10^{10} cm/s

Energy is generally expressed in electron volts (eV) (1 eV = 1.602×10^{-12} erg).

Reference

Sawyer, C. N., P. L. McCarty, and G. F. Parkin. 1994. *Chemistry for Environmental Engineering*, McGraw-Hill, New York.

ELECTRICAL DOUBLE LAYER: Thickness

Introduction

In aquatic environments, organic and inorganic colloidal particles as well as biocolloids (e.g., viruses, bacteria) are generally negatively charged. In the presence of an electrolyte, the particles attract a cloud of positively charged ions that are organized in two layers (Figure E2): (1) the *Stern layer*, a compact layer of counter ions adjacent to the charged colloid surface; and (2) the *Gouy–Chapman diffuse double layer*, containing the rest of the counter ions, which extend in the bulk of the solution (Stumm and Morgan, 1981; Sundstrom and Klei, 1979; Bitton and Marshall, 1980). The thickness of the electrical double layer decreases with an increase in salt concentration and valence of the cation.

Formulas

$$z = 0.33 \times 10^{-2}\left(\frac{\varepsilon}{I}\right)^{1/2}$$

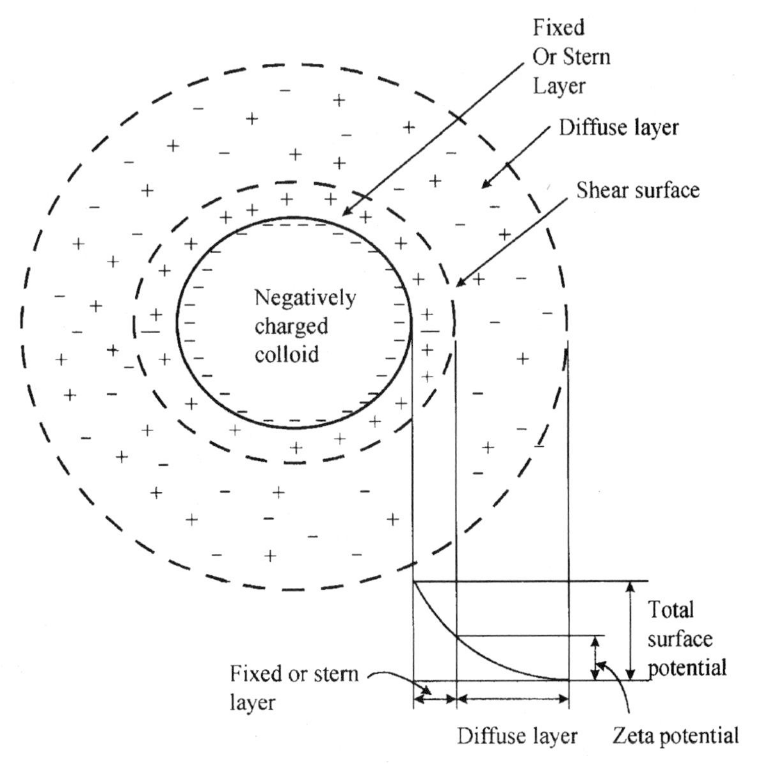

FIGURE E2: Electrical Double Layer of a Negatively Charged Colloidal Particle (*Source:* Sawyer et al. 1994.)

where

z = thickness of double layer (cm)

ε = dielectric constant of solution (C/V · cm)

I = ionic strength of solution (mol/L)

For water at 20°C, the thickness is

$$z = 2.8 \times 10^{-8} I^{-1/2}$$

The double-layer thickness is conventionally reported as $1/\kappa$. For clay surfaces, κ is given by the following (McBride, 1994):

$$\kappa = Az\left(\frac{n_0}{\varepsilon kT}\right)^{1/2}$$

where

A = constant
z = counterion charge
n_0 = electrolyte concentration
ε = dielectric constant of the solvent
k = Boltzmann constant
T = absolute temperature (K)

The formula above shows that an expanded double layer is obtained by using a counter ion with low charge (e.g., Na^+), a solution with low salt concentration, or a solvent with high dielectric constant (McBride, 1994).

References

Bitton, G., and K. C. Marshall, Eds. 1980. *Adsorption of Microorganisms to Surfaces*, Wiley, New York.

McBride, M. B. 1994. *Environmental Chemistry of Soils*, Oxford University Press, New York.

Sawyer, C. N., P. L. McCarty, and G. F. Parkin. 1994. *Chemistry for Environmental Engineering*, McGraw-Hill, New York.

Stumm, W., and J. J. Morgan. 1981. *Aquatic Chemistry*, Wiley-Interscience, New York.

Sundstrom, D. W., and H. E. Klei. 1979. *Wastewater Treatment*, Prentice Hall, Upper Saddle River, NJ.

ELECTROPHORETIC MOBILITY: Colloidal Particles

Definition

Electrophoretic mobility of charged colloidal particles is the velocity of particles over a known distance in an electric field.

Formula

$$\bar{u} = \frac{u c_s q}{I}$$

where

$\bar{u}$ = electrophoretic mobility of particle (e.g., bacteria, virus) (mm/s · V · cm)
u = particle velocity (mm/s)
c_s = specific conductivity of buffer solution (mS/cm)

q = cross-sectional area of electrophoresis cell (cm^2)
i = current (A)

References

Abramson, H. A., L. S. Moyer, and M. H. Gorin. 1942. *Electrophoresis of Proteins and the Chemistry of Cell Surfaces*, Reinhold, New York.

Marshall, K. C. 1976. *Interfaces in Microbial Ecology*, Harvard University Press, Cambridge, MA.

ENERGY OF A PHOTON

Definition

The energy, E, of a photon or light quanta is related to the wavelength of light by the formula that follows.

Formula

$$E = \frac{hc}{\lambda}$$

where

E = energy of photon (J)
h = Planck's constant = 6.625×10^{-34} J · s
c = speed of light = 3×10^8 m/s
λ = wavelength of light (m)

Thus ultraviolet rays have higher energy than microwaves or radio waves, which have higher wavelengths.

Reference

McElroy, W. D. 1971. *Cell Physiology and Biochemistry*, 3rd ed., Prentice-Hall, Upper Saddle River, NJ.

ENZYME ACTIVITY

Definition

Enzyme activity is the amount of substrate converted or the amount of product formed per unit time under optimal assay conditions.

Formulas

$$\text{Enzyme activity} = \frac{\text{amount of substrate converted}}{t}$$

or

$$= \frac{\text{amount of product formed}}{t}$$

Enzyme activity can be expressed in international units (IU) or in katal (kat).

- 1 IU = 1 μmol substrate converted/min. It is the amount of enzyme that converts 1 μmol of substrate per minute.
- 1 kat = 1 mol substrate/s = 6.0×10^7 IU.

Reference

Efiok, B. J. S. 1993. *Basic Calculations for Chemical and Biological Analyses.* AOAC International, Arlington, VA.

ENZYME INHIBITION: Competitive Inhibition of Enzymatic Reactions

Introduction

In competitive inhibition, an inhibitor (I) and the enzyme substrate (S) compete for the same site on the enzyme (E).

Formula

In the presence of a competitive inhibitor I, the reaction rate V of a given enzyme is given by

$$V = \frac{V_{\max}[\mathrm{S}]}{[\mathrm{S}] + K_m(1 + [\mathrm{I}]/K_i)}$$

where

$V_{\max}$ = maximum specific rate (1/time)

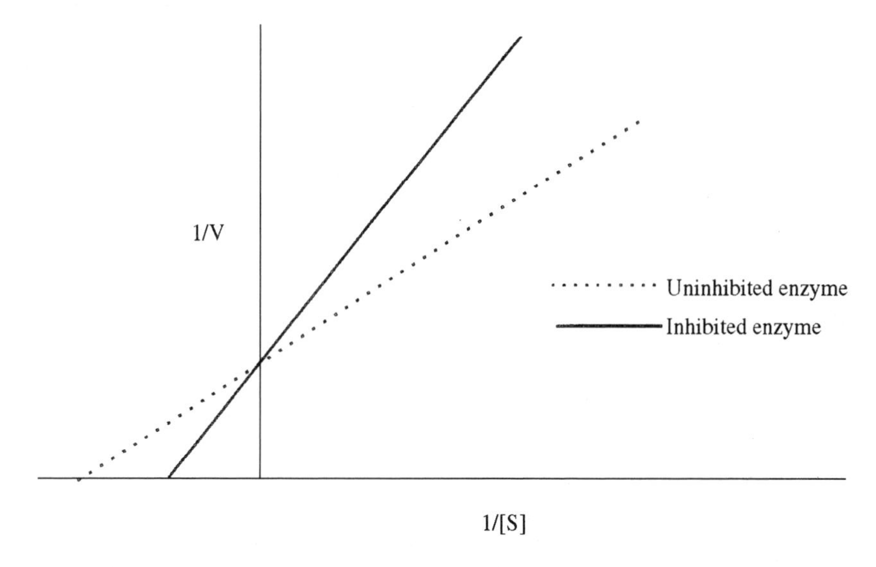

FIGURE E3: Lineweaver–Burk Plot for Competitive Inhibition of Enzymes (*Source:* Clark and Switzer, 1977.)

[S] = substrate concentration (mol/L)
K_m = Michaelis constant (mol/L)
[I] = inhibitor concentration (mol/L)
K_i = inhibition constant (mol/L)

The reciprocal form of the equation above is

$$\frac{1}{V} = \frac{K_m}{V_{\max}}\left(1 + \frac{[\mathrm{I}]}{K_i}\right)\frac{1}{[\mathrm{S}]} + \frac{1}{V_{\max}}$$

A plot of $1/V$ versus 1/[S] is given in Figure E3. The slope increases as [I] increases. However, the $1/V$ intercept remains $1/V_{\max}$.

References

Clark, J. M., and R. L. Switzer. 1977. *Experimental Biochemistry*, W.H. Freeman, San Francisco.

Grady, C. P. L., and H. C. Lim. 1980. *Biological Wastewater Treatment: Theory and Applications*, Marcel Dekker, New York.

Segel, I. H. 1976. *Biochemical Calculations*, 2nd ed., Wiley, New York.

ENZYME INHIBITION: Noncompetitive Inhibition of Enzymatic Reactions

Introduction

In noncompetitive enzyme inhibition, the inhibitor I may combine with either the enzyme E or with the complex enzyme–substrate ES.

Formulas

$$V = \frac{V_{\max}[S]}{(K_m + [S])(1 + [I]/K_i)}$$

Using the Lineweaver–Burk transformation, the reaction rate V is given by

$$\frac{1}{V} = \frac{1}{V_{\max}}\left(1 + \frac{[I]}{K_i}\right) + \frac{K_m}{V_{\max}}\left(1 + \frac{[I]}{K_i}\right)\frac{1}{[S]}$$

The double reciprocal plots for noncompetitive inhibition are shown Figure E4. In the presence of the inhibitor, the slope of the double reciprocal plot is increased, whereas $V_{\max}$ is decreased. K_m remains unchanged.

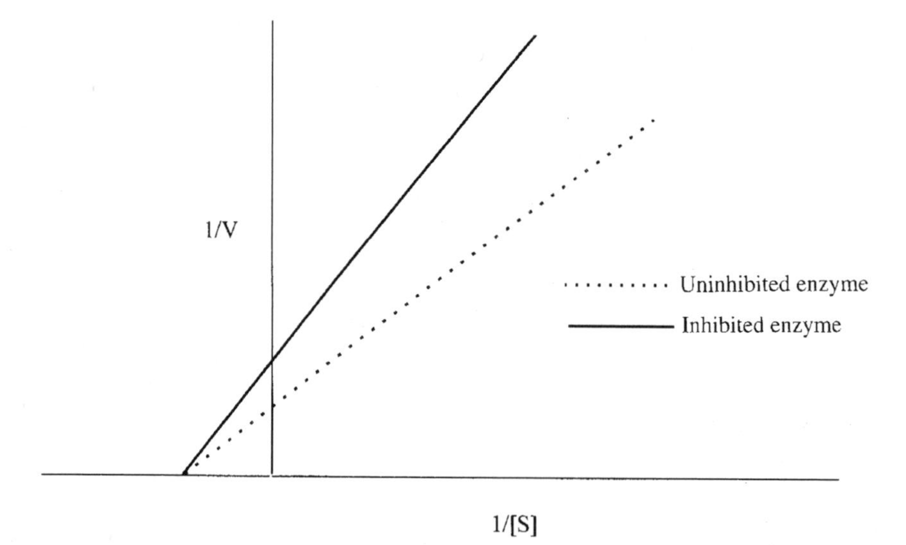

FIGURE E4: Lineweaver–Burk Plot for Noncompetitive Inhibition of Enzymes (*Source:* Clark and Switzer, 1977.)

References

Clark, J. M., and R. L. Switzer. 1977. *Experimental Biochemistry*, W.H. Freeman, San Francisco.

Michal, G. 1978. Determination of Michaelis constants and inhibitor constants, pp. 29–40. In: *Principles of Enzymatic Analysis*, H. U. Bergmeyer, Ed., Verlag Chemie, Weinheim, Germany.

Segel, I. H. 1976. *Biochemical Calculations*, 2nd ed., Wiley, New York.

ENZYME INHIBITION: Uncompetitive Inhibition of Enzymatic Reactions

Definition

In incompetitive enzyme inhibition, the inhibitor binds to the enzyme–substrate complex (ES) but not to the free enzyme.

Formula

In the presence of an uncompetitive inhibitor I, the reaction rate V is given by

$$V = \frac{V_{max}[S]}{K_m + [S](1 + [I]/K_i)}$$

where

V_{max} = maximum specific rate (1/time)
[S] = substrate concentration (mol/L)
K_m = Michaelis constant (mol/L)
[I] = inhibitor concentration (mol/L)
K_i = inhibition coefficient

The Lineweaver–Burk plot transformation gives the equation

$$\frac{1}{V} = \frac{K_m}{V_{max}}\frac{1}{[S]} + \frac{1}{V_{max}}\left(1 + \frac{[I]}{K_i}\right)$$

In the presence of an inhibitor, the slope of the reciprocal plot remains the same but both V_{max} and K_m are affected by the inhibitor concentration (Figure E5).

References

Clark, J. M., and R. L. Switzer. 1977. *Experimental Biochemistry*, W.H. Freeman, San Francisco.

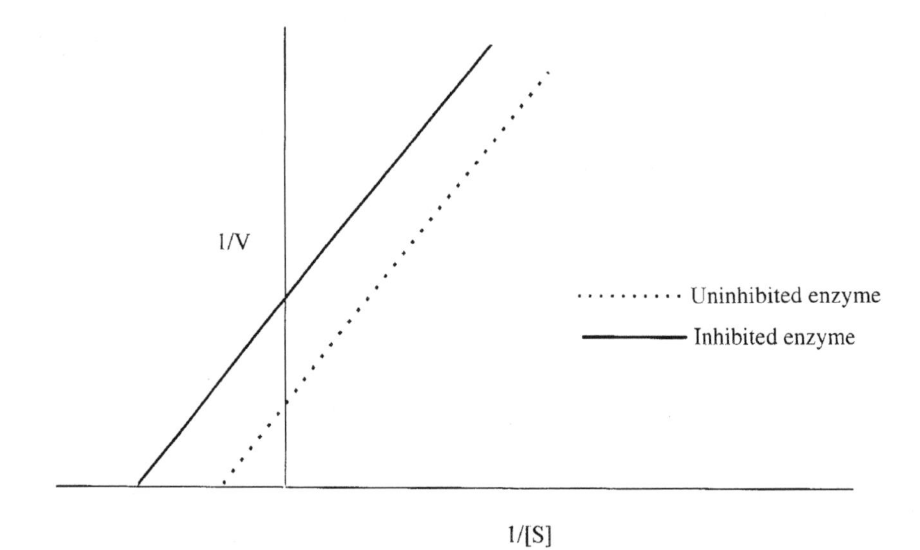

FIGURE E5: Lineweaver–Burk Plot for Unconcompetitive Inhibition of Enzymes (*Source:* Clark and Switzer, 1977.)

Grady, C. P. L., and H. C. Lim. 1980. *Biological Wastewater Treatment: Theory and Applications*, Marcel Dekker, New York.

Michal, G. 1978. Determination of Michaelis constants and inhibitor constants, pp. 29–40. In: *Principles of Enzymatic Analysis*, H. U. Bergmeyer, Ed., Verlag Chemie, Weinheim, Germany.

Segel, I. H. 1976. *Biochemical Calculations*, 2nd ed., Wiley, New York.

ENZYME KINETICS

See DIXON PLOTS; EADIE–SCATCHARD PLOT; ENZYME INHIBITION; HANES EQUATION; HOFSTEE EQUATION; LINEWEAVER–BURK EQUATION; MICHAELIS–MENTEN EQUATION.

ENZYMES: Specific Activity

Definition

The specific activity of an enzyme is the amount of enzyme activity measured as the amount of substrates (S) converted or product (P) formed per unit time *t* per amount of protein present under specified assay conditions.

Formula

$$\text{Specific activity} = \frac{\text{enzyme units}}{\text{mg protein}}$$

where enzymes units expressed as μmol substrate used/min or as μmol product formed/min.

Reference

Clark, J. M., and R. L. Switzer. 1977. *Experimental Biochemistry*, 2nd ed., W.H. Freeman, San Francisco.

ENZYMES: Turnover number or Catalytic Constant

Definition

Turnover number is the maximum number of moles of substrate S converted per mole of active site on an enzyme per second.

Formula

$$\text{Turnover number} = \frac{\text{moles of S converted}}{\text{s} \times \text{moles of enzyme}}$$

Reference

Efiok, B. J. S. 1993. *Basic Calculations for Chemical and Biological Analyses*, AOAC International, Arlington, VA.

EQUITABILITY (or Evenness)

Definition

In terms of the structure of communities, the *equitability* or *evenness* E is defined as the ratio between the observed diversity index H and the maximum theoretical diversity index H_{max}. H_{max} is calculated by assuming an equal number of individuals in all species.

Formula

$$E = \frac{\bar{H}}{\bar{H}_{\max}}$$

Reference

Ramade, F. 1981. *Ecology of Natural Resources*, Wiley, New York.

EXPONENTIAL GROWTH OF MICROORGANISMS

Introduction

In a batch culture, microbial populations undergo a lag phase, an exponential phase, a stationary phase, and a death phase (Figure E6).

Formula

During the exponential growth phase, microbial growth is given by the equation

$$\frac{dX}{dt} = \mu X \Rightarrow X = X_0 e^{\mu t}$$

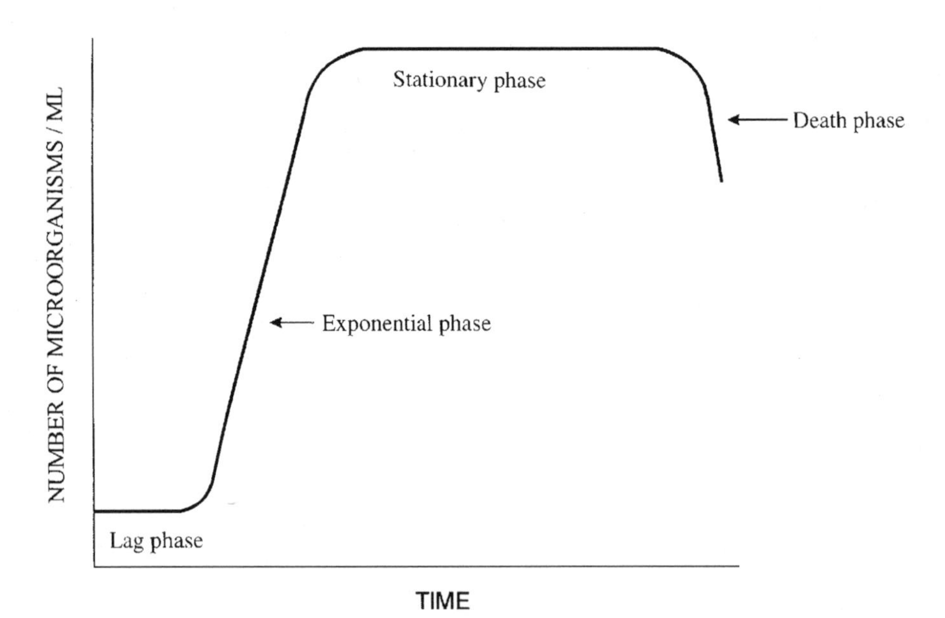

FIGURE E6: Growth Curve for Microorganisms.

where

X = biomass concentration (g/L or cell numbers) at time t

μ = specific growth rate (h^{-1})

X_0 = biomass concentration (g/L or cell numbers) at time = 0

t = time (h)

The doubling time t_d of the culture is given by

$$t_d = \frac{\ln 2}{\mu} = \frac{0.693}{\mu}$$

Reference

Drew, S. W. 1981. Liquid cultures, pp. 151–178. In: *Manual of Methods for General Bacteriology*, P. Gerhardt et al., Eds., American Society for Microbiology, Washington, DC.

F

FECAL COLIFORMS/FECAL STREPTOCOCCI (FC/FS) RATIO

Introduction

The relative quantities of fecal coliforms and fecal stretococci discharged by humans are different from the relative quantities discharged by animals. Thus the fecal coliform to fecal streptococci ratio is indicative of the source of fecal contamination (human versus animal).

Numerical Values

See Table F1.

Comments:

1. The sample pH should be between 4 and 9 to exclude any adverse effects of pH on either group of microorganisms.
2. At least two counts should be made on each sample.
3. To minimize errors due to differential death rates, samples should not be taken farther downstream than 24 hours of flow time from the suspected source of contamination.

Reference

Mara, D. D. 1974. *Bacteriology for Salinity Engineers*, Churchill Livingstone, Edinburgh.

TABLE F1: Typical FC/FS Ratios

Source	FC/FS
Chicken	0.4
Cow	0.2
Duck	0.6
Pig	0.04
Sheep	0.4
Turkey	0.1
Human	4.4

FICK'S LAW

Introduction

The first Fick's law (1855) relates to the diffusion of a substrate through a stagnant liquid film or biofilm. Diffusion is the net movement of chemicals from levels of high to low concentrations and is determined by the diffusion coefficient of the solute and its concentration gradient (Koch, 1990; Williams et al., 1978).

Formulas

$$J = -D\frac{dC}{dx}$$

where

J = substrate mass flux in the x direction per unit area per unit time ($\text{mol/cm}^2 \cdot \text{s}$)

D = molecular diffusion coefficient ($\text{cm}^2\text{/s}$)

C = substrate concentration (mol/cm^3)

dC/dx = concentration gradient (mol/cm^4)

It is assumed that the temperature and viscosity remain constant.

The diffusion coefficient D, determined by measuring the diffusion through a porous glass disk, is given by the following (Tinoco et al., 1995):

$$D = -J\frac{\Delta x}{c_2 - c_1}$$

where

J = substrate mass flux ($\text{mol/cm}^2 \cdot \text{s}$)

Δx = thickness of the porous glass disk (cm)

c_2, c_1 = chemical concentration on both sides of the porous glass disk (mol/cm^3)

Fick's law can also be written as follows:

$$J = k_L(C_b - C_s)$$

where

J = substrate flux ($\text{mol/cm}^2 \cdot \text{s}$)

k_L = mass transfer coefficient (cm/s)

C_b = substrate concentration in bulk phase (mol/cm^3)

C_s = substrate concentration at the surface (mol/cm^3)

References

Grady, C. P. L., and H. C. Lim. 1980. *Biological Wastewater Treatment: Theory and Applications*, Marcel Dekker, New York.

Koch, A. L. 1990. Diffusion: the crucial process in many aspects of the biology of bacteria. Adv. Microb. Ecol. 37–70.

Schwarzenbach, R. P., P. M. Gschwend, and D. M. Imboden. 1993. *Environmental Organic Chemistry*, Wiley-Interscience, New York.

Tinoco, I., Jr., K. Sauer, and J. C. Wang. 1995. *Physical Chemistry: Principles and Applications in Biological Sciences*, Prentice Hall, Upper Saddle River, NJ.

Williams, V. R., W. L. Mattice, and H. B. Williams. 1978. *Basic Physical Chemistry for the Life Sciences*, W.H. Freeman, San Francisco.

FILTER COEFFICIENT

Definition

In sand filtration, the filter coefficient λ gives the rate of change of particle (e.g., mineral or organic particle, bacterial cells) concentration with respect to filter depth.

Formulas

The particle concentration profile throughout a sand filter is given by

$$\frac{dC}{dz} = -\lambda C$$

where

C = particle concentration (number/L)
z = depth (m)
λ = filter coefficient (m^{-1})

In bacterial transport though sand columns, bacterial numbers are determined before and after passage though the sand column. The filter coefficient is given by

$$\lambda = \frac{\ln(C/C_0)}{L}$$

where

C = outlet bacterial concentration
C_0 = inlet bacterial concentration
L = column length (m)

Reference

Tien, C. 1989. *Granular Filtration of Aerosols and Hydrosols*, Butterworth, Newton, MA.

FILTERING RATE OF ZOOPLANKTON

See CLEARANCE RATE: Zooplankton and Protozoa.

FLOC LOAD

See ACTIVATED SLUDGE: Floc Load

FLOW RATE: Domestic Wastewater

Introduction

The flow rate of domestic wastewater on a per capita basis varies with the population. In small systems (population less than 10,000), a general relationship has been suggested. At populations of 10,000 or more, a constant per capita contribution of 0.38 m^3/day is typically assumed.

Formula

$$\frac{Q}{P} = 3.8 \times 10^{-3}(50 + P/200)$$

where

Q = average daily wastewater flow, m^3/day
P = population served ($0 < P < 10{,}000$)

Reference

Rich, L. G. 1980. *Low-Maintenance, Mechanically Simple Wastewater Treatment Systems*, McGraw-Hill, New York.

FOOD-TO-MICROORGANISM (F/M) RATIO

Also called *process loading factor*.

Definition

F/M ratio is the ratio between the organic loading rate to an activated sludge system and the mass of sludge in the system.

Formula

Organic loading rate is expressed in terms of biochemical oxygen demand (*see* BIOCHEMICAL OXYGEN DEMAND) or chemical oxygen demand. Sludge mass is expressed in terms of total dry weight (*see* MIXED LIQUOR SUSPENDED SOLIDS) or ash-free dry weight (*see* MIXED LIQUOR VOLATILE SUSPENDED SOLIDS). The F/M ratio indicates the organic load into the activated sludge system and is given by (Curds and Hawkes, 1983; Nathanson, 1986)

$$F/M = \frac{Q \times \text{BOD}}{\text{MLVSS} \times V}$$

where

F/M = kg BOD/kg MLVSS per day
Q = flow rate of sewage (m^3/day)
BOD = 5-day biochemical oxygen demand in influent (mg/L)
MLVSS = mixed liquor volatile suspended solids (mg/L)
V = volume of aeration tank (m^3)

Numerical Values

F/M is controlled by the rate of activated sludge wasting. The higher the wasting rate, the higher the F/M ratio. For conventional aeration tanks the F/M ratio is

TABLE F2: Food to Microorganisms Ratios in Some Activated Sludge Systems

Process	F/M Ratio (kg BOD/kg MLVSS · day)
Conventional	0.2–0.4
Step Aeration	0.2–0.4
Contact Stabilization	0.2–06
Extended aeration	0.05–0.15
Pure oxygen system	Up to 1.5

Source: Data from Hammer (1986), Metcalf and Eddy (1991).

0.2–0.5 kg BOD_5/kg MLVSS · day but can be higher (up to 1.5) for activated sludge using high-purity oxygen (Hammer, 1986) (Table F2). A low F/M ratio means that the microorganisms in the aeration tank are starved, leading to a more efficient wastewater treatment.

References

Curds, C. R., and H. A. Hawkes, Eds. 1983. *Ecological Aspects of Used-Water Treatment*, vol. 2, Academic Press, London.

Hammer, M. J. 1986. *Water and Wastewater Technology*, Wiley, New York.

Metcalf and Eddy, Inc. 1991. *Wastewater Engineering: Treatment, Disposal and Reuse*, 3rd ed., McGraw-Hill, New York.

Nathanson, J. A. 1986. *Basic Environmental Technology: Water Supply, Waste Disposal and Pollution Control*, Wiley, New York.

FREE ENERGY

Definition/Introduction

Free energy is the energy available for doing work (e.g., macromolecule synthesis) in a biochemical reaction. In biochemical reactions, a portion of energy released is lost as a result of increase in entropy. The rest of the energy is useful energy or free energy.

Formula

The change in free energy is given by

$$\Delta G = \Delta H - T\Delta S$$

where

ΔG = Gibbs free energy of a chemical reaction (kcal/mol)

ΔH = enthalpy or heat of reaction (kcal/mol)

T = absolute temperature (K)

ΔS = entropy (kcal/K · mol)

Chemical reactions that release free energy have a negative ΔG. They proceed from reactants to products and are called *exergonic reactions*. Chemical reactions that require an input of energy and proceed from products to reactants, and have a positive ΔG are called *endergonic reactions*. Figure F1 shows the changes in free energy during exergonic and endergonic reactions (Atlas, 1986).

References

Atlas, R. M. 1984. *Basic and Practical Microbiology*, Macmillan, New York.

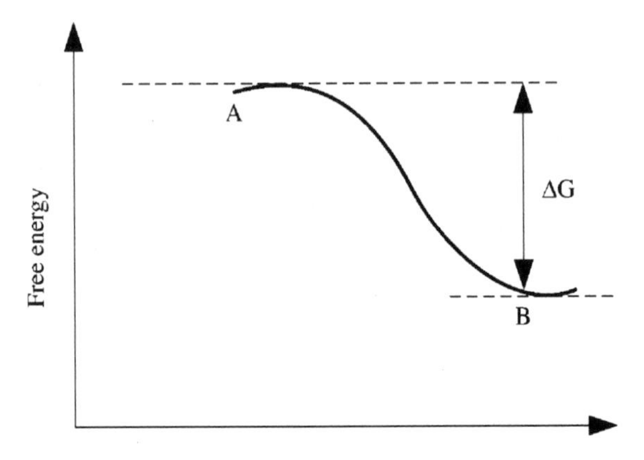

FIGURE F1: Changes in Free Energy During Exergonic and Endergonic Reactions.

Ray, B. T. 1995. *Environmental Engineering*, PWS, Boston.

Williams, V. R., W. L. Mattice, and H. B. Williams. 1978. *Basic Physical Chemistry for the Life Sciences*, 3rd ed., W.H. Freeman, San Francisco.

FREE ENERGY: Changes Accompanying Chemical Reactions

See also STANDARD FREE ENERGY.

Introduction

Chemical reactions result in changes in free energy (ΔG). The direction of the chemical reaction depends on the concentrations of chemical reactants and products. Let us consider the following chemical reaction:

$$aA + bB \rightleftharpoons c\mathrm{C} + d\mathrm{D}$$

where A and B are the reactants and C and D are the products of the chemical reaction.

Formula

The free-energy change accompanying this chemical reaction is

$$\Delta G = \Delta G^0 + RT \ln \frac{[C]^c [D]^d}{[A]^a [B]^b}$$

where

ΔG = free energy (J)
ΔG_0 = standard free energy (J)
R = universal gas constant = 8.314 J/K · mol
T = absolute temperature (K)
[A],[B] = activities or effective concentrations of reactants
[C],[D] = activities or effective concentrations of products

At equilibrium, ΔG is equal to zero and ΔG^0 is

$$\Delta G_0 = \frac{[C]^c[D]^d}{[A]^a[B]^b}$$
$$= -RT \ln K_{eq}$$

where K_{eq} is the equilibrium constant.

Numerical Values

See Table F3

TABLE F3: Free Energies of Hydrolysis of Some Phosphorylated Compounds

Compound	ΔG near 25°C (kcal/mol)
Phosphoenolpyruvate	−14.8
1,3-Diphosphoglycerate	−11.8
Creatine phosphate	−10.5
Acetyl phosphate	−10.3
Inorganic pyrophosphate	−8.0
Arginine phosphate	−7.7
ATP	−7.3
Glucose-1-phosphate	−5.0
Glucose-6-phosphate	−3.0
Fructose-1-phosphate	−3.1
3-Phosphoglycerate	−3.1

Source: Data from Lehninger (1971), Sober (1970), Clark and Switzer (1977).

References

Clark, J. M., and R. L. Switzer, 1977. *Experimental Biochemistry*, W.H. Freeman, San Francisco.

Gaudy, A. F. Jr., and E. T. Gaudy. 1988. *Elements of Bioenvironmental Engineering*, Engineering Press, San Jose, CA.

Lehninger, A. L. 1971. *Bioenergetics*, 2nd ed., W. A. Benjamin, Menlo Park, CA.

Sawyer, C. N., and P. L. McCarty. 1978. *Chemistry for Environmental Engineering*, McGraw-Hill, New York.

Sober, H. R., Ed. *Handbook of Biochemistry*, 2nd ed., CRC Press, Boca Raton, FL, p. J-185.

Williams, V. R., W. L. Mattice, and H. B. Williams, 1978. *Basic Physical Chemistry for the Life Sciences*, 3rd ed., W.H. Freeman, San Francisco.

FRESNEL'S FORMULA

See LIGHT REFLECTION.

FREUNDLICH ISOTHERM

Definition/Introduction

The empirically derived Freundlich isotherm describes the partitioning of a chemical compound or a microorganism between liquid and solid compartments or phases. It is commonly used to describe adsorption of contaminants onto sorbents such as activated carbon or the adsorption of microbial pathogens or indicators onto soils or aquifer materials.

Formulas and Numerical Values

$$q_e = K_f C^{1/n}$$

where

q_e $= \frac{x}{m} =$ amount of adsorbate x (chemical contaminant or microorganisms) adsorbed per unit weight m of adsorbent.

C = concentration of compound or microorganisms in liquid compartment or phase.

K_f and n = empirical constants

TABLE F4: Freundlich Isotherm: Values of K_f and $1/n$ for Some Priority Organic Pollutants at Neutral pH

Organic Pollutant	K_f (mg/g)	$1/n$
Chlorobenzene	93	0.98
Dibromochloromethane	63	0.93
Hexachlorobutadiene	360	0.63
Hydroquinone	90	0.25
α-Naphtol	180	0.31
Nitrobenzene	68	0.43
Pentachlorophenol	150	0.42
p-Xylene	85	0.16

Source: Adapted from Eckenfelder (1989).

K_f is the equilibrium constant of chemical or microorganism and indicates the strength of adsorption; $1/n$ is the slope of the isotherm. These constants depends on the type of absorbent and the type of substance adsorbed. Table F4 gives K_f and $1/n$ for some priority pollutants.

The linearized form of the Freundlich isotherm is given by

$$\log q_e = \log K_f + \frac{1}{n} \log C$$

Plotting log q_e versus log C gives a straight line illustrated in Figure F2. K_f and n are determined from the plot (the slope of the line is $1/n$ and the y-intercept is log K_f).

As regards microbial adsorption to soils, K_f depends on several factors, which include soil and microorganism types, pH, chemical composition of the suspending fluid, and soluble organic materials. For example, for poliovirus type 1, K_f varied from 0.72 to 1000 (mL/g), depending on the soil type and the suspending medium (Powelson and Gerba, 1995). For some viruses, $1/n$ ranged between 0.87 and 1.24 (Viker and Burge, 1980).

The estimated *retardation factor* R_e (retardation factor measures the transport, though a soil or a sand column, of a microorganism relative to the bulk of the water) for a given microorganism is related to the constant K_f by the following equation (Powelson and Gerba, 1995):

$$R_e = 1 + \frac{\rho_b K_f}{\theta_v}$$

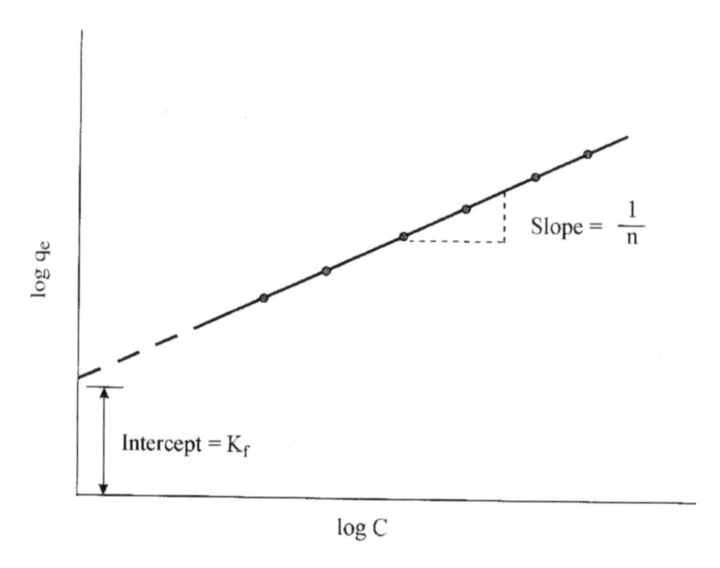

FIGURE F2: Plot of Freundlich Isotherm.

where

ρ_b = soil bulk density (g/cm^3)

θ_v = volumetric water content (cm^3)

References

Benefield, L. D., J. F. Judkins, Jr., and B. L. Weand. 1982. *Process Chemistry for Water and Wastewater Treatment*, Prentice Hall, Upper Saddle River, NJ.

Eckenfelder, W. W. Jr., 1989. *Industrial Water Pollution Control*, 2nd ed., McGraw-Hill, New York.

Freundlich, H. 1926. *Colloid and Capillary Chemistry*, Methuen, London.

Powelson, D. K., and C. P. Gerba. 1995. Fate and transport of microorganisms in the vadose zone, pp. 123–135. In: *Handbook of Vadose Zone Characterization and Monitoring*, L. G. Wilson, L. G. Everett, and S. J. Cullen, Eds., Lewis Publishers, Boca Raton, FL.

Shane, B. S. 1994. Principles of ecotoxicology, pp. 11–47. In: *Basic Environmental Toxicology*, L. G. Cockerman and B. S. Shane, Eds., CRC Press, Boca Raton, FL.

Shaw, D. J. 1966. *Introduction to Colloid and Surface Chemistry*, Butterworth, London.

Vilker, V. L., and W. D. Burge. 1980. Adsorption mass transfer model for virus transport in soils. Water REs. 14: 783–790.

Weber, W. J. Jr. 1972. *Physicochemical Processes for Water Quality Control*, Wiley-Interscience, New York.

G

GAS LAW: Ideal Gas Law or Perfect Gas Law

Definition

The gas law (ideal gas law or ideal gas equation) relates to the effect of temperature, pressure, and the number of moles on gas volume.

Formula

$$PV = nRT$$

where

P = pressure (atm)
V = volume (L)
n = number of moles of gas in a given sample
R = universal gas constant = $0.082\,\mathrm{L \cdot atm/mol \cdot K} = 8.2\,\mathrm{J/mol \cdot K}$
T = temperature (K)

References

Battley, E. H. 1987. *Energetics of Microbial Growth*, Wiley, New York.

Sawyer, C. N., and P. L. McCarty. 1978. *Chemistry for Environmental Engineering*, McGraw-Hill, New York.

Wanielista, M. P., A. Y. Yousek, J. S. Taylor, and C. D. Cooper. 1984. *Engineering and the Environment*, Brooks/Cole, Pacific Grove, CA.

GAS PRODUCTION IN FACULTATIVE PONDS

Introduction

Biological solids from wastewater settle to the bottom of facultative ponds, forming a sludge blanket. The biological solids in the blanket will undergo anaerobic degradation at a rate that is temperature dependent. Solid conversion to gas as a function of temperature is given by the following:

Formulas

$$G = 6.28 \times 10^{-3}(T_s - 15)$$

where

G = solids converted to gas (kg BOD/m$^2\cdot$day)
T_s = sludge temperature (°C)

Sludge temperature may be predicted by the following formula:

$$T_s = 0.777T_a + 0.89$$

where T_a is the monthly mean maximum air temperature at the water surface.

Reference

Oswald, W. G. 1970. Designing waste ponds to meet water quality criteria. Proc. 2nd Int. Symp. for Waste Treatment Lagoons, Kansas City, MO.

GENERATION TIME OF MICROORGANISMS

See DOUBLING TIME OF MICROORGANISMS.

GOUY–CHAPMAN DIFFUSE DOUBLE LAYER

See ELECTRICAL DOUBLE LAYER.

GROWTH: Competition for a Single Growth-Limiting Substrate

Introduction

In an open system, assuming Monod's kinetics, the competition between two organisms (organisms 1 and 2) for the same growth-limiting substrate with a concentration S is described by the following equations, which show the growth of both organisms as well as the change of substrate concentration with time.

Formulas

- *Organism 1*

$$\frac{dX_1}{dt} = \frac{\mu_{max1} S X_1}{K_{s1} + S} - DX_1$$

- *Organism 2*

$$\frac{dX_2}{dt} = \frac{u_{max2} S X_2}{K_{s2} + S} - DX_2$$

- *Substrate concentration*

$$\frac{dS}{dt} = D(S_i - S) - \frac{\mu_{max1} S X_1}{Y_1(K_{s1} + S)} - \frac{\mu_{max2} S X_2}{Y_2(K_{s2} + S)}$$

where

X_1 = concentration of organism 1
X_2 = concentration of organism 2
S = substrate concentration (mg/L)
μ_{max1} = maximum specific growth rate of organism 1 (h^{-1})
μ_{max2} = maximum specific growth rate of organism 2 (h^{-1})
K_{s1} = half saturation constant for organism 1 (mg/L)
K_{s2} = half saturation constant for organism 2 (mg/L)
D = dilution rate (h^{-1})
Y_1 = yield coefficient for organism 1
Y_2 = yield coefficient for organism 2

Reference

Bazin, M., and A. Menell. 1990. Mathematical methods in microbial ecology. Methods Microbiol. 22: 125–179 (R. Grigorova and J. R. Norris, Eds.), Academic Press, London.

GROWTH: Microbial Colonies

Introduction

Microorganisms inoculated into a solid growth medium form colonies that grow until the nutrients are depleted or until they become inhibited by toxic metabolites.

Formulas

Colony growth or cell mass is described by the following:

$$\frac{dM}{dt} = \mu M$$

or

$$\ln M_t = \ln M_0 + \mu t$$

where

M_0 = original cell mass (g)
M_t = cell mass at time t (g)
μ = specific growth rate (h^{-1})

The mass of a circular colony that grows to a height h with a radius r is given by

$$M = \pi r^2 h \rho$$

where ρ is the colony density (g/cm^3).

From the equations above, we derive the following relationship between colony radius and specific growth rate μ:

$$\ln r = \frac{\mu}{2} t + \ln r_0$$

where r_0 is the radius of the original cell at time = 0. Thus the slope of the plot of ln r versus t is linearly related to the specific growth rate μ.

Reference

Sikyta, B. 1995. *Techniques in Applied Microbiology*, Elsevier, Amsterdam.

GROWTH: Surface Colonization Rate

See SURFACE MICROBIAL COLONIZATION EQUATION.

GROWTH EFFICIENCY

See ECOLOGICAL EFFICIENCIES.

GROWTH, POPULATION

Introduction

The growth of a population in a closed environment when resources or environmental factors are not limiting is given by the *J-shaped population curve* (Figure G1).

Formulas

$$\frac{dX}{dt} = rX$$

where

r = rate of increase of the population in the absence of competition (1/time)

X = number of individuals in the population (population size) at a given time

This can also be expressed as

$$X_t = X_0 e^{rt}$$

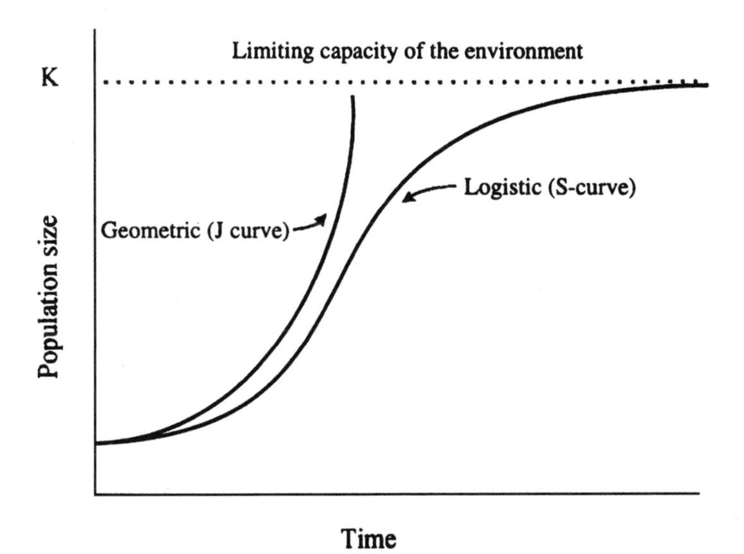

FIGURE G1: Theoretical Population Growth in the Absence (J Curve) and Presence (S Curve) of Limiting Factors (*Source:* Adapted from Ramade, 1981.)

where

X_0 = number of individuals in the population at time 0

X_t = number of individuals in the population at time t

If the environment is limited (e.g., limited resources), the population growth is slowed down and reaches the maximum number the environment can support. This maximum number is called the *carrying capacity of the environment K*. Under these conditions, population growth can be described by the *logistic curve* (i.e., sigmoid or S-shaped curve) first described by the French mathematician P. F. Verhulst in 1838 and later derived independently by the Americans R. Pearl and L. J. Reed in 1920 to describe the growth of the US population. The growth of a population under these conditions is shown by the S-shaped curve in Figure G1 (Ramade, 1981; Bazin and Menell, 1990).

$$\frac{dX}{dt} = rX\left(\frac{K - X}{K}\right)$$

where

r = rate of increase of the population (1/time)

X = number of individuals in the population at given time

K = carrying capacity of the environment (or maximal value of X)

The population growth will decrease as X increases to approach the carrying capacity K and will be equal to zero as $X = K$. The r value is influenced by physical (e.g., temperature, light, humidity) and chemical (nutrient concentration, trace element concentration, toxic chemicals) factors (Jorgensen and Johnsen, 1989).

References

Bazin, M., and A. Menell. 1990. Mathematical methods in microbial ecology. Methods Microbiol. 22: 125–179 (R. Grigorova and J. R. Norris, Eds.), Academic Press, London.

Jorgensen, S. E., and I. Johnsen. 1989. *Principles of Environmental Science and Technology*, Elsevier, Amsterdam.

Kormondy, E. J. 1996. *Concepts of Ecology*, 4th ed., Prentice Hall, Upper Saddle River, NJ.

Krebs, C. J. 1972. *Ecology*, Harper & Row, New York.

Pearl, R., and L. J. Reed. 1920. On the rate of growth of the population of the United States since 1790 and its mathematical representation. Proc. Natl. Acad. Sci. USA 6: 275–288.

Ramade, F. 1981. *Ecology of Natural Resources*, Wiley, New York.

GROWTH RATE OF MICROORGANISMS: Minimum Substrate Concentration (S_{min})

Introduction

Microorganisms grow well at substrate concentrations close to the saturation constant K_s. However, for a given microorganism–substrate combination, growth is not possible when the substrate concentration reaches a threshold limit S_{min}.

Formula

$$S_{min} = b \frac{K_s}{\mu_{max}}$$

where

S_{min} = minimum or threshold substrate concentration (mg/L)
b = decay rate of microorganisms (h^{-1})
μ_{max} = maximum specific growth rate (h^{-1})

Numerical Values

b ranges between 0.005 and 0.02 h^{-1}. For drinking water microorganisms, b is assumed to be 0.01 h^{-1}, giving an S_{min} of 0.1 to 0.2 μg C/L (van der Kooij, 1995).

Reference

van der Kooij, D. 1995. Significance and assessment of the biological stability of drinking water, pp. 89–102. In: *Water Pollution: Quality and Treatment of Drinking Water*, Springer Verlag, New York.

GROWTH RATE OF MICROORGANISMS: Observed Rates of Increase of Microbial Populations

Introduction

In open systems where microorganisms (e.g., bacteria, algae) are continuously supplied with substrates and also continuously removed by grazers, the observed rate of increase of the population is the result of growth and predation.

Formula

$$r = \mu - m$$

where

r = observed rate of increase of population (h^{-1})

μ = specific growth rate (h^{-1})

m = removal rate by grazers (h^{-1})

$m = F\langle D\rangle$

where

F = clearance rate (volume/grazer per unit of time)

$\langle D\rangle$ = time-averaged density of grazers (number of grazers/volume)

References

Ducklow, H. W., and S. M. Mill. 1985. The growth of heterotrophic bacteria in the surface waters of warm core rings. Limnol. Oceanogr. 30: 239–259.

Landry, M. R. 1993. Estimating rates of growth and grazing mortality of phytoplankton by the dulution method. In: *Handbook of Methods in Aquatic Microbial Ecology*, P. F. Kemp, B. F. Sherr, E. B. Sherr, and J. J. Fisher, Eds., Lewis Publishers, Boca Raton, FL.

GROWTH RATE OF MICROORGANISMS: Relationship of Growth with Substrate Concentration

See MONOD'S EQUATION.

GROWTH YIELD OF MICROORGANISMS

Definition

Growth yield is the amount of biomass produced per unit amount of substrate removed. It reflects the efficiency of conversion of substrates to cell biomass.

Formulas

$$\text{Yield } Y = \frac{\text{mass of dells produced}}{\text{mass of substrate used}} = \frac{\Delta X}{\Delta S} = \frac{X - X_0}{S_0 - S}$$

where

S_0, S = initial and final substrate concentrations, respectively (g/L or mol/L)
X_0, X = initial and final microbial concentrations, respectively (g/L)

Several factors influence the growth yield and are the following: type of microorganisms, growth medium, substrate concentration, terminal electron acceptor, pH, and incubation temperature. Growth yield Y is related to the specific growth rate μ and substrate uptake rate q by the following equation:

$$\mu = Yq - k_d$$

where

μ = specific growth rate (h^{-1})
q = substrate uptake rate (h^{-1})
k_d = endogeneous decay rate (h^{-1})

Numerical Values

Under laboratory conditions, bacteria have growth yields ranging from 0.4 to 0.6 g biomass/g substrate (Horan, 1989; Rittmann, 1995) (Table G1). Growth yield of bacteria ranges from 0.1 to 0.4 in environmental samples (Lancelot and Billen, 1984, 1985; Linley et al., 1983; Newell et al., 1981, 1983; Servais et al., 1987). The growth yield of nutrifying bacteria in activated sludge at 15°C was reported as $Y = 0.2$ (Horan, 1989).

Growth Yield of Protozoa Finlay (1993) gave the following formula for calculating the growth yield (also known as *gross growth efficiency*) of a ciliated protozoan (the prey can be bacterial or yeast cells):

$$Y = \frac{\mu V_c}{U V_p}$$

TABLE G1: Growth Yields Y of Some Microorganisms

Microorganism (Substrate)	Y (g cells/g substrate)
Escherichia coli (complex medium)	0.66
Nitrifiers (g cells/g N) (ammonium)	0.33
Denitrifiers (acetate)	0.27
Phosphate-accumulating bacteria (acetate)	0.5–0.6
Saccharomyces cerevisiae (alcoholic fermentation)	0.11

where

μ = specific growth rate of the ciliate ($\mu = \ln 2/T_d$ where T_d = doubling time of the population)

V_c = average ciliate cell volume (μm^3)

U = rate of uptake of prey cells by the ciliate (cells/ciliate · h)

V_p = average prey cell volume (μm^3)

Following is an example of an anaerobic marine ciliate, *Metopus contortus*, feeding on yeast cells (Finlay, 1993):

$$\mu = 0.0092 \text{ h}^{-1}$$

$$U = 98 \text{ yeast cells/ciliate} \cdot \text{h}$$

$$V_c = 1.3 \times 10^5 \ \mu\text{m}^3$$

$$V_p = 135 \ \mu\text{m}^3$$

Then $Y = 0.09$ (9% in terms of volume).

References

Drew, S. W. 1981. Liquid culture, pp. 151–178. In: *Manual of Methods for General Bacteriology*, P. Gerhardt et al., Eds., American Society for Microbiology, Washington, DC.

Finlay, B. J. 1993. Behavior and bioenergetics of anaerobic and microaerobic protists, pp. 59–66. In: *Handbook of Methods in Aquatic Microbial Ecology*, P. F. Kemp, B. F. Sherr, E. B. Sherr, J. Cole, Eds., Lewis Publishers, Boca Raton, FL.

Henze, M., P. Harremoës, J. La Cour Jansen, and E. Arvin. 1995. *Wastewater Treatment: Biological and Chemical Processes*, Springer-Verlag, Berlin.

Horan, N. J. 1989. *Biological Wastewater Treatment Systems: Theory and Operation*, Wiley, New York.

Lancelot, C., and G. Billen. 1984. Activity of heterotrophic bacteria and its coupling to primary production during the spring phytoplankton bloom in the Southern Bight of the North Sea. Limnol. Oceanogr. 29: 721–730.

Lancelot, C., and G. Billen. 1985. Carbon–nitrogen relationships in the dynamics of coastal marine ecosystems. Adv. Aquat. Microbiol. 3: 263–321.

Linley, E. A., R. C. Newell, and M. I. Lucas. 1983. Quantitative relationships between phytoplankton, bacteria and heterotrophic microflagellates in shelf waters. Mar. Ecol. Prog. Ser. 12: 77–89.

Newell, R. C., M. I. Lucas, and E. A. Linley. 1981. Rate of degradation and efficiency of conversion of phytoplankton debris by marine microorganisms. Mar. Ecol. Prog. Ser. 6: 123–136.

Newell, R. C., E. A. Linley, and M. I. Lucas. 1983. Bacterial production and carbon conversion based on salt marsh plant debris. Estuarine Coastal Shelf Sci. 17: 405–420.

Rittmann, B. E. 1995. Fundamentals and application of biofilm processes in drinking-water treatment, pp. 61–87. In: *Water Pollution: Quality and Treatment of Drinking Water*, Springer-Verlag, New York.

Servais, P., G. Billen, and M.-C. Hascoet. 1987. Determination of the biodegradable fraction of dissolved organic matter in waters. Water Res. 21: 445–450.

Shamat, N. A., and W. J. Maier. 1980. Kinetics of biodegradation of chlorinated organics. J. Water Pollut. Control Fed. 52: 2158–2166.

Stanier, R. Y., L. L. Ingraham, M. L. Wheelis, and P. R. Painter. 1986. *The Microbial World*, Prentice Hall, Upper Saddle River, NJ.

H

HALDANE EQUATION

Introduction

When relatively toxic compounds are used as substrates for microbial growth, the specific growth rate increases with increasing substrate concentration up to a critical substrate concentration. Beyond that critical level, the specific growth rates decrease as a result of substrate toxicity.

Formula

$$\mu = \frac{\mu_{max} S}{K_s + S + S^2/K_i}$$

where

μ = specific growth rate (h^{-1})
μ_{max} = maximum specific growth rate (h^{-1})
S = substrate concentration (mg/L)
K_s = saturation constant (mg/L)
K_i = inhibition constant (mg/L)

Figure H1 shows the specific growth rate as a function of the substrate concentration in the presence of a nontoxic substrate (Monod's relationship) and in the presence of a toxicant (Haldane relationship).

References

Kennedy, M. S., J. Grammas, and W. B. Arbuckle. 1990. Parachlorophenol degradation using bioaugmentation. J. Water Pollut. Control Fed. 62: 227–233.

Rozich, A. F., and A. F. Gaudy, Jr. 1985. Response of phenol-acclimated activated sludge process to quantitative shock loading. J. Water Pollut. Control Fed. 57: 795–804.

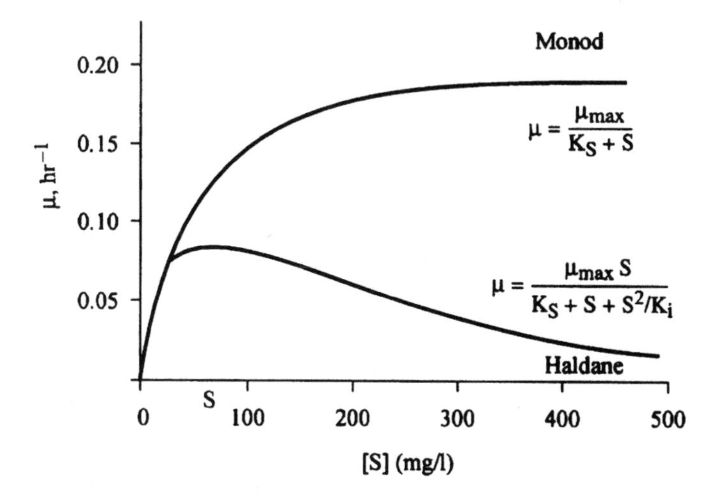

FIGURE H1: Haldane Equation: Comparison of the Monod and Haldane Plots (*Source:* Rozich and Gaudy, Jr., 1985.)

HALF-LIFE OF CHEMICALS ($t_{1/2}$)

Definition

The half-life of a chemical is the length of time required for 50% of a chemical to react (disappear).

Formula

	zero-order reaction	*first-order reaction*	*second-order reaction*
	A → P	A → P	2A → P
$t_{1/2} =$	$\frac{0.5A_0}{k}$	$\frac{0.693}{k}$	$\frac{1}{A_0 k}$

where

A_0 = initial concentration of component A

k = rate constant:

zero-order (mass/time)

first-order (1/time)

second-order [1/mass · time)

Reference

Wanielista, M. P., Y. A. Yousef, J. S. Taylor, and C. D. Cooper. 1984. *Engineering and the Environment*, Brooks/Cole, Pacific Grove, CA.

HANES EQUATION

Also called the Hanes–Woolf equation.

Introduction

The *Hanes equation* is a linearized form of the Michaelis–Menten equation. It may be used to estimate values for the parameters V_{max} and K_m from experimental data.

Formula

$$\frac{[S]}{V} = \frac{1}{V_{max}}[S] + \frac{K_m}{V_{max}}$$

where

- $[S]$ = substrate concentration (mol/L)
- V = reaction rate (1/time)
- V_{max} = maximum reaction rate (1/time)
- K_m = half saturation constant (Michaelis constant; the substrate concentration at which V is equal to $V_{max}/2$)

A plot of $[S]/V$ versus [S] is linear and has a slope of $1/V_{max}$, an intercept of K_m/V_{max} on the y-axis, and an intercept of $-K_m$ on the x-axis (Figure H2).

References

Hanes, C. S. 1932. Studies on plant amylases. I. The effect of starch concentration upon the velocity of hydrolysis by the amylase of germinated barley. Biochem. J. 26: 1406–1421.

Segel, I. H. 1976. *Biochemical Calculations*, 2nd ed., Wiley, New York.

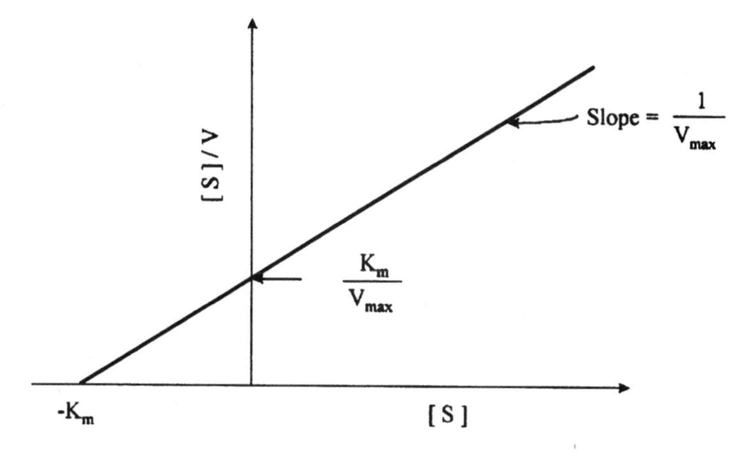

FIGURE H2: Hanes Plot (*Source:* Adapted from Segel, 1976.)

HARDNESS, TOTAL

Definition

Total hardness is the sum of calcium and magnesium ion concentrations in water. It is expressed as mg/L of $CaCO_3$. Other cations causing hardness are Sr^{2+}, Fe^{2+}, and Mn^{2+}.

Numerical Values

In the United States, waters are classified as soft (0–60 mg/L $CaCO_3$), moderately hard (60–120 mg/L $CaCO_3$), hard (120–180 mg/L $CaCO_3$), and very hard (>180 mg/L $CaCO_3$).

Reference

Benefield, L. D., J. F. Judkins, Jr., and B. L. Weand. 1982. *Process Chemistry for Water and Wastewater Treatment*, Prentice Hall, Upper Saddle River, NJ.

HELMOTZ–SMOLUCHOWSKI EQUATION

Introduction

The Helmotz–Smoluchowski equation gives the relationship between the electrophoretic mobility of a colloidal particle (e.g., bacterial cell) and its zeta potential.

Formula

$$\zeta = \frac{4\pi h\bar{u}}{D}$$

where

ζ = zeta potential (mV)
h = viscosity (cP)
$\bar{u}$ = electrophoretic mobility of particle (e.g., bacteria, virus) (mm/s · V · cm)
D = dielectric constant

In water at 25°C, $\zeta = 12.9\bar{u}$.

Reference

Marshall, K. C. 1976. *Interfaces in Microbial Ecology*, Harvard University Press, Cambridge, MA.

HENDERSON–HASSELBALCH EQUATION

Introduction

The Henderson–Hasselbalch equation describes the degree of ionization of an acid or base.

Formulas

$$\text{Acids:} \qquad \mathrm{pK}_a = \mathrm{pH} + \log\frac{[\text{nonionized}]}{[\text{ionized}]}$$

$$\text{Bases:} \qquad \mathrm{pK}_a = \mathrm{pH} + \log\frac{[\text{ionized}]}{[\text{nonionized}]}$$

where pK_a is the pH at which 50% of a compound is ionized and 50% is nonionized.

In toxicology, the degree of ionization of a chemical is useful for predicting the passage of drugs and toxic chemicals through biological membranes. Non-ionized compounds are better absorbed than ionized compounds. Organic acids are better absorbed in acidic environments (e.g., stomach).

References

Lehninger, A. L. 1973. *Short Course in Biochemistry*, Worth Publishers, New York.

Shane, B. S. 1994. Principles of ecotoxicology, pp. 11–47. In: *Basic Environmental Toxicology*, L. G. Cockerman and B. S. Shane, Eds., CRC Press, Boca Raton, FL.

HENRY'S LAW CONSTANT

Definition

Henry's law constant K_H is an expression of the equilibrium distribution of a compound between water and gas phase (air) and is the ratio of the concentration of a compound in the gas phase to its concentration in the aqueous phase at equilibrium. It is indicative of the volatility of a compound.

Formula

$$K_H = \frac{P_i}{C_w}$$

where

K_H = Henry's law constant ($atm \cdot m^3/mol$)
P_i = partial pressure of the compound in the gas phase (atm)
C_w = compound concentration in the aqueous phase (mol/m^3)

Henry's law constant is temperature dependent, as organic compounds volatilazation decreases with temperature. This constant gives an indication of the removal of organics by air stripping in engineering reactors (Sawyer et al., 1994).

Numerical Values

Table H1 displays the Henry's constant for some organic compounds.

References

Sawyer, C. N., P. L. McCarty, and G. F. Parkin. 1994. *Chemistry for Environmental Engineering*, McGraw-Hill, New York.

Schnoor, J. L., et al. 1987. *Processes, Coefficients, and Models for Simulating Toxic Organics and Heavy Metals in Surface Waters*, EPA 600/3-87/015, U.S. Environmental Protection Agency, Washington, DC.

Schwarzenbach, R. P., P. M. Gschwend, and D. M. Imboden. 1993. *Environmental Organic Chemistry*, Wiley-Interscience, New York.

Thibodeaux, L. J. 1979. *Chemodynamics*, Wiley, New York.

TABLE H1: Henry's Law Constants for some Organic Compounds

Compound	K_H (atm · m^3/mol at 20°C)
Halogenated aliphatics	
Vinyl chloride	2.4
1,1-Dichloroethylene	0.352
Chloroethane	0.148
1,2,-*trans*-Dichloro-ethylene	0.067
Carbon tetrachloride	0.023
1,1,1-Trichloroethane	0.018
Tetrachloroethylene	0.012
Trichloroethylene	0.0088
1,1-Dichloroethane	0.0043
Chloroform	0.0032
Dichloromethane	0.002
1,2-Dichloroethane	0.0011
1,1,2-Trichloroethane	0.00074
Bromoform	0.0005
Aromatics	
Ethylbenzene	0.0088
Toluene	0.0066
Benzene	0.0055
Ethyl benzene	0.0087
Chlorobenzene	0.0037
1,3-Dichlorobenzene	0.0036
1,2-Dichlorobenzene	0.0018
Hexachlorobenzene	0.00068
Naphthalene	4.6×10^{-4}
Anthracene	1.65×10^{-3}
Phenanthrene	1.48×10^{-4}
Nitrobenzene	2.2×10^{-5}
2-Chlorophenol	1.03×10^{-5}
2-Nitrophenol	7.56×10^{-6}
Pentachlorophenol	3.4×10^{-6}
Benzo[a]pyrene	4.9×10^{-7}
Phenol	4.57×10^{-7}
Miscellaneous organics	
DDT	3.8×10^{-5}
Lindane	4.8×10^{-7}
Dieldrin	2.0×10^{-7}
2,4-D	1.72×10^{-9}

Source: Data from Schnoor et al. (1987), Thibodeaux (1979).

HETEROTROPHIC POTENTIAL

Introduction

Heterotrophic potential measures the activity of natural heterotrophic microbial populations in aquatic environments. To do so, one studies the uptake of radiolabeled compounds (e.g., glucose, glutamate, acetate). Dissolved organic compounds occur at *low concentrations* in natural waters (μg/L). For example, dissolved free amino acids are 25–100 μg/L in estuarine and coastal waters and 2–10 μg/L in offshore waters.

Formulas

One of the approaches used to measure heterotrophic potential is the V_{max}–kinetics approach. It uses a series of concentrations of a labeled substrate (μg/L range). The uptake of organic substrates by heterogeneous microbial populations can be described by Michaelis–Menten enzyme kinetics.

$$V = \frac{f}{t}(S_n + A) \tag{1}$$

where

V = uptake rate (μg C/L · h)
f = fraction of added isotope taken up
S_n = natural concentration of substrate (μg/L)
A = added substrate (μg/L)

The Michaelis–Menten equation is

$$V = V_{max}\frac{S}{K_s + S} \tag{2}$$

where

V_{max} = maximum uptake rate (μg C/L · h)
K_s = half saturation constant (μg/L)
S = substrate concentration = $S_n + A$(μg/L)

By combining equations (1) and (2), we obtain

$$\frac{t}{f} = \frac{1}{V_{max}}A + \frac{K_t + S_n}{V_{max}} \tag{3}$$

where

t = incubation time (h)

f = fraction of isotope taken up

A = concentration of substrate added (μg/L)

This modified Lineweaver–Burk plot of Michaelis–Menten kinetics give the plot shown in Figure H3. The plot gives the following parameters: Slope = $1/V_{max}$, which gives the value of V_{max} = maximum uptake velocity (μg/L · h) and T_t = turnover time (h).

This approach is applicable to coastal waters, lakes, and sediments. V_{max} is the full capacity substrate uptake rate. It was proposed to use it as an *indicator of heterotrophic potential.* V_{max} for assimilation is obtained if $^{14}CO_2$ production is measured. T_t, the turnover time, is the time for the bacteria to utilize all existing substrate.

Numerical Values

$$V_{max} = 10^{-4}\text{–}10 \ \mu g/L \cdot h$$

$T_t = 0.5$ h in polluted pond

$T_t = 7\text{–}93$ h in estuaries (for amino acids)

$T_t = 5000$ h in an oligotrophic lake

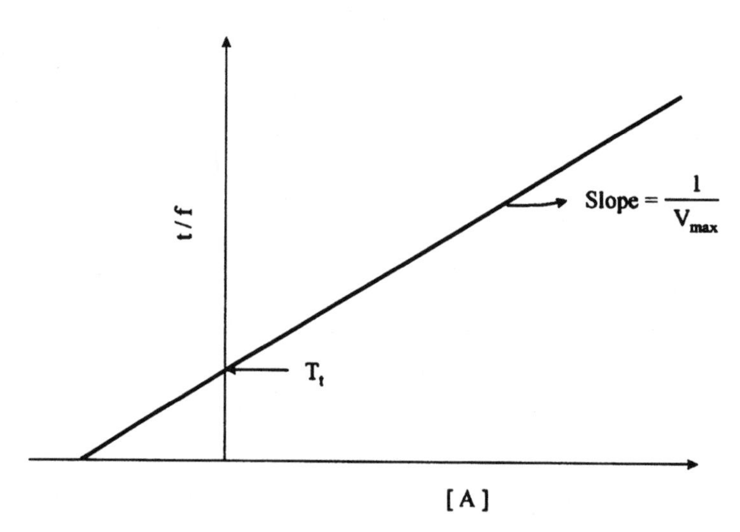

FIGURE H3: Lineweaver–Burk Plot for Measuring Heterotrophic Potential.

References

Hobbie, J. E., and C. C. Crawford. 1969a. Bacterial uptake of organic substrates: new methods of study and application to eutrophication. Verh. Int. Ver. Limnol. 17: 725–730.

Hobbie, J. E., and C. C. Crawford. 1969b. Respiration corrections for bacterial uptake of dissolved organic compounds in natural waters. Limnol. Oceanogr. 14: 528–532.

Wright, R. T. 1978. Measurement and significance of specific activity in the heterotrophic bacteria in natural waters. Appl. Environ. Microbiol. 36: 297–305.

Wright, R. T., and B. K. Burnison. 1979. Heterotrophic activity measured with radiolabeled organic substrates, pp. 140–155. In: *Native Aquatic Bacteria: Enumeration, Activity and Ecology*. STP 695, W. Costerton and R. Colwell, Eds. American Society for Testing Materials, Philadelphia.

Wright, R. T., and J. E. Hobbie. 1966. Use of glucose and acetate by bacteria and algae in aquatic ecosystems. Ecology 47: 447–464.

HOFSTEE EQUATION

Also called the Woolf–Augustinsson–Hofstee plot. *See also* MICHAELIS–MENTEN EQUATION.

Definition

The Hofstee equation is a linearized form of the Michaelis–Menten equation. It may be used to estimate values for the kinetic parameters V_{max} and K_m.

Formula

$$V = -K_m \frac{V}{[S]} + V_{max}$$

where

V = reaction rate (1/time)
K_m = half saturation constant (Michaelis constant; the substrate concentration at which V is equal to $V_{max}/2$)
$[S]$ = substrate concentration (mol/L)
V_{max} = maximum reaction rate (1/time)

The plot of V versus $V/[S]$ is linear with a slope of $-K_m$, a y-intercept of V_{max}, and an x-intercept of V_{max}/K_m (Figure H4).

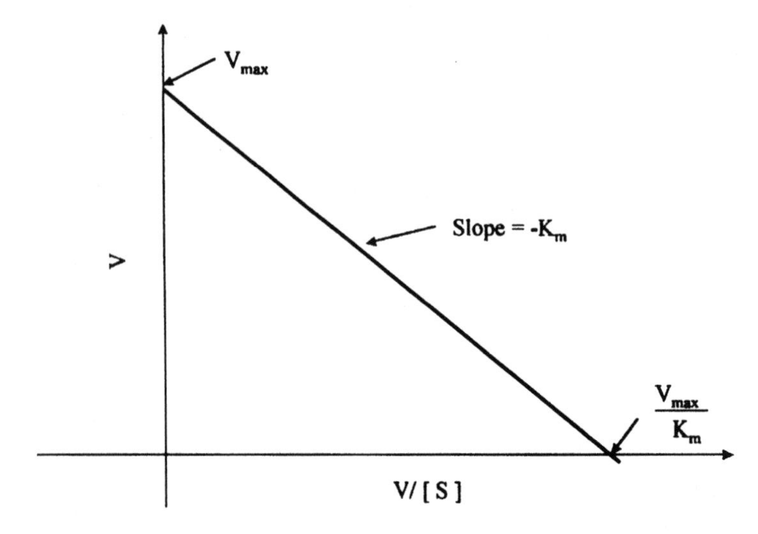

FIGURE H4: Hofstee Plot (*Source:* Adapted from Segel, 1976.)

Reference

Segel, I. H. 1976. *Biochemical Calculations*, 2nd ed., Wiley, New York.

HYDRAULIC CONDUCTIVITY

See DARCY'S LAW.

HYDRAULIC RETENTION TIME (HRT)

Definition

Hydraulic retention time represents the time that a liquid stays in a reactor.

Formula

$$\text{HRT} = \frac{V}{Q}$$

where

HRT = hydraulic residence time (days)
V = reactor volume (m^3)
Q = flow rate (m^3/day)

Reference

Sundstrom, D. W., and H. E. Klei. 1979. *Wastewater Treatment*, Prentice Hall, Upper Saddle River, NJ.

HYDROGEN ION ACTIVITY

See pH.

HYDROPHOBICITY: Cell Surface Hydrophobicity Determination

Introduction

Microbial cell surface hydrophobicity can be determined by measuring the distribution of microbial cells in a two-phase system consisting of an aqueous buffer solution and an organic solvent (e.g., heptane or hexadecane). Two volumes of cells suspended in a buffer are added to 1 volume of organic solvent and mixed for 20 s. The optical density of the buffer solution is measured at time 0 and after 30 min.

Formula

Cell surface hydrophobicity (CSH) is given by

$$\mathrm{CSH}\ (\%) = 100\left(\mathrm{OD_i} - \frac{\mathrm{OD_{30}}}{\mathrm{OD_i}}\right)$$

where

OD_i = initial optical density of the buffer solution

OD_{30} = optical density of the buffer solution after 30 min

References

Rosenberg, M., D. Gutnick, and E. Rosenberg. 1980. Adherence of bacteria to hydrocarbons: A simple method for measuring cell-surface hydrophobicity. FEMS Microbiol. Lett. 9: 29–33.

Suzzi, G,. P. Romano, and L. Vannini. 1994. Cell surface hydrophobicity and flocculence in *Saccharomyces cerevisiae* wine yeasts. Colloids Surf. B Biointerfaces 2: 505–510.

HYDROSTATIC PRESSURE

Definition/Introduction

In aquatic environments, hydrostatic pressure is the pressure exerted by a stratum of water at a given depth. Hydrostatic pressure, in addition to temperature, atmospheric pressure, and water salinity, influences the amount of gas (e.g., oxygen) that is dissolved in water. This, in turn, affects metabolic activity in aquatic environments.

Formula

Hydrostatic pressure is related to the pressure at the surface of the water by the following formula:

$$P_z = P_0 + 0.0967z$$

where

P_z = hydrostatic pressure at depth z (atm)
P_0 = atmospheric pressure at the surface (atm)
z = water depth (m)

Reference

Wetzel, R. G. 1975. *Limnology*, W. B. Saunders, Philadelphia.

I

INACTIVATION RATE, MICROBIAL

See CHICK'S LAW.

INGESTION RATE: Zooplankton or Protozoa

Definition

The ingestion rate of zooplankton or protozoa is the amount of prey (bacteria or algae) ingested per unit time.

Formulas

$$I = \frac{F_t - F_0}{Xt}$$

where

I = ingestion rate expressed as number of ingested algal or bacterial cells/animal per day
F_t = prey concentration at time t (cells/mL)
F_0 = prey concentration at time 0 (cells/mL)
X = number of animals (predators)
t = time

Ingestion rate I is related to clearance rate CR by the following equation (Porter et al., 1982):

$$I = \mathrm{CR} \times F$$

where

CR = clearance rate (mL/animal per unit of time)
F = prey concentration in feeding suspension (cells/mL)

Numerical Values

- Maximum ingestion rate for the dinoflagellete *Protoperidinium hirobis* feeding on diatoms (batch cultures) = 23 diatoms/dinoflagellate per day.
- Using gut fluorescence (fluorescence due to feeding on algae) as an indicator of feeding by a marine cladoceran (*Penilia avirostris*) in Hong Kong harbor, the following mean ingestion rates I were reported (Wong et al., 1992):

  June: $I = 43.1$ ng pigment/animal per day (range = 9.2–101.4) (28 stations investigated).

  November: $I = 26.3$ ng pigment/animal per day (range = 13.8–50.7) (28 stations investigated).
- Laboratory studies with *Monas* sp. feeding on bacteria (*Escherichia coli*, *Salmonella typhimurium*, *Chlorobium phaeobacterioides*, and an unidentified bacterial isolate from Lake Kinneret, Israel): The ingestion rates varied between 10 and 75 bacteria/microflagellate per hour, corresponding to 30–200% of the average microflagellate body weight per hour. The feeding rate increased with the food concentration (Sherr at al., 1983).

References

Porter, K. G., J. Gerritsen, and J. D. Orcutt, Jr. 1982. The effect of food concentration on swimming patterns, feeding behavior, ingestion, assimilation, and respiration by *Daphnia*. Limnol. Oceanogr. 27: 935–949.

Sherr, B. F., E. B. Sherr, and T. Berman. 1983. Grazing, growth, and ammonium excretion rates of a heterotrophic microflagellate fed with four species of bacteria. Appl. Environ. Microbiol. 45: 1196–1201.

Wong, C. K., A. L. Chan, and K. W. Tang. 1992. Natural ingestion rates and grazing impact of the marine cladoceran *Penilia avirostris* Dana in Tolo harbour, Hong Kong. J. Plankton Res. 14: 1757–1765.

IONIC STRENGTH

Definition

Ionic strength I is a parameter that expresses the concentration of all ionic solutes in water, in terms of their molarity and their charge (Williams et al., 1978).

Formulas

Ionic strength is one-half the sum of the concentrations of all ions. Each concentration is multiplied by the square of the electronic charge of the ion:

$$I = \frac{1}{2}\sum C_i Z_i^2$$

where

I = ionic strength (conventionally dimensionless)

C_i = ion concentration of the ith species (mol/L)

Z_i = charge on the ith species

- *Relationship of ionic strength to total dissolved solids (TDS) in water and wastewater* (Benefield et al., 1982): For aquatic environments for which TDS is less than 1000 mg/L, the relationship between TDS and I is given by (Langelier, 1936)

$$I = (2.5 \times 10^{-5})(\text{TDS})$$

Another relationship between I and TDS is given by Kemp (1971) to account for nonionic silica, which contributes to TDS:

$$I = (2.5 \times 10^{-5})(\text{TDS} - 20)$$

- *Relationship between I and electrical conductance* (Kemp, 1971):

$$I = (2.5 \times 10^{-5})(\text{EC})(g)$$

where

EC = electrical conductance (S/cm)

g = proportionality factor = 0.55–0.70 [a value of $g = 0.67$ is used; sometimes g is outside this range (Benefield et al., 1982)]

References

Benefield, L. B., J. F. Judkins, Jr., and B. L. Weand. 1982. *Process Chemistry for Water and Wastewater Treatment*, Prentice Hall, Upper Saddle River, NJ.

Kemp, P. H. 1971. Chemistry of natural waters. Water Res. 5: 297–304.

Langelier, W. F. 1936. Effect of temperature on the pH of natural waters. J. Am. Water Works Assoc. 28: 1500–1504.

Williams, V. R., W. L. Mattice, and H. B. Williams. 1978. *Basic Physical Chemistry for the Life Sciences*, 3rd ed., W.H. Freeman, San Francisco.

ION PRODUCT OF WATER

Definition

K_w, the ion product of water, is the product of the H^+ and OH^- concentrations in water and is equal to 10^{-14} at 25°C.

Formula

$$K_w = [H^+][OH^-]$$

where

K_w = ion product of water
$[H^+]$ = hydrogen ion concentration
$[OH^-]$ = hydroxide ion concentration

Reference

Lehninger, A. L. 1973. *Short Course in Biochemistry*, Worth Publishers, New York.

L

LAGOON TEMPERATURE

Formula

The following formula enables the estimation of the temperature of a wastewater lagoon, based on lagoon size, wastewater flow rate, and air temperature.

$$T_w = \frac{AfT_a + QT_i}{Af + Q}$$

where

T_w = lagoon water temperature (°C)
A = lagoon surface area
f = proportionality factor
T_a = ambient air temperature (°C)
Q = wastewater flow rate (Liter/time unit)
T_i = influent wastewater temperature (°C)

Reference

Mancini, J. L., and E. L. Barnhart. 1968. Industrial waste treatment in industrial aerated lagoons. In: *Advances in Water Quality Improvements*, E. F. Gloyna and W. W. Eckenfelder, Eds., University of Texas Press, Austin, TX.

LANGMUIR ISOTHERM

See also ADSORPTION ISOTHERMS.

Introduction

The Langmuir isotherm describes the adsorption of contaminants onto adsorbents such as activated carbon. The Langmuir isotherm assumes the absorption of a single layer of adsorbate on the surface of the adsorbent. It is also assumed that all the adsorption sites have equal affinity for the molecules and that the adsorption at one site does not affect adsorption at an adjacent site (Langmuir, 1918; Weber, 1972).

Formula

$$q_e = \frac{Q^0 b C_e}{1 + bC_e}$$

where

q_e = amount of adsorbate adsorbed per unit weight of adsorbent (g/g)
Q^0 = number of moles of solute/unit weight of adsorbent to form a monolayer on the surface (mass/mass)
b = empirical constant related to the energy of adsorption (L/mg)
C_e = equilibrium concentration of adsorbate in solution after adsorption (mg/L)

The Langmuir isotherm is shown in Figure L1a.

A linearized form of the Langmuir isotherm is given by the equation

$$\frac{1}{q_e} = \frac{1}{Q^0} + \frac{1}{bQ^0}\left(\frac{1}{C_e}\right)$$

A plot of $1/q_e$ versus $1/C_e$ gives a straight line with a slope of $1/bQ^0$ and a y-intercept of Q^0 (see Figure L1b).

References

Langmuir, I. 1918. The adsorption of gases on plane surfaces of glass, mica and platinum. J. Am. Chem. Soc. 40: 1361–1367.

Shaw, D. J. 1966. *Introduction to Colloid and Surface Chemistry*, Butterworth, London.

Weber, W. J., Jr. 1972. *Physicochemical Processes for Water Quality Control*, Wiley-Interscience, New York.

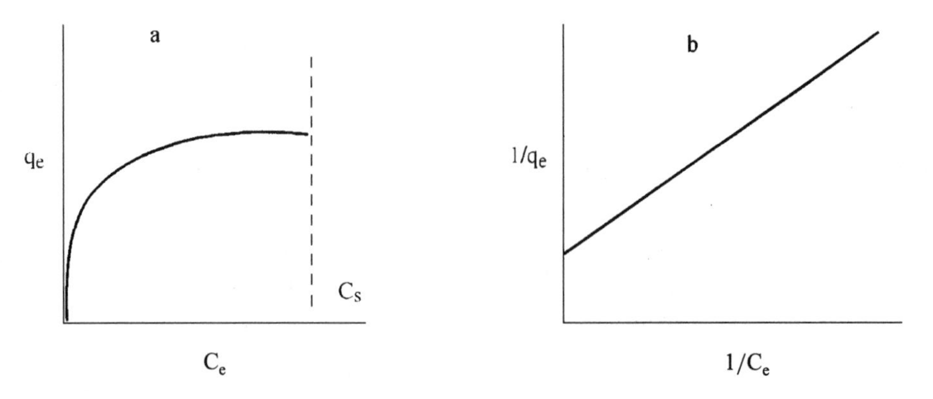

FIGURE L1: Langmuir Isotherms: (a) Typical; (b) Linear (*Source:* Weber, 1972.)

LETHALITY COEFFICIENT

See DISINFECTION.

LIEBIG'S LAW: Law of the Minimum

Introduction

This law was formulated by Liebig in the nineteenth century. According to Liebig's law, the biomass or yield of an organism (e.g., plant, animal, microorganism) is controlled by the nutrient (e.g., nitrogen, phosphorus) that is present at the lowest (minimum) concentration. At the other extreme of the concentration scale, excessive levels of an element decrease or prevent the growth of an organism.

References

Atlas, R. M., and R. Bartha. 1981. *Microbial Ecology: Fundamentals and Applications*, Addison-Wesley, Reading, MA.

Liebig, J. 1840. *Chemistry in Its Application to Agriculture and Physiology*, Taylor and Walton, London.

Ramade, F. 1981. *Ecology of Natural Resources*, Wiley, New York.

LIGHT: Compensation Light Intensity

Definition

In lakes and other aquatic environments, light intensity decreases exponentially with depth. The compensation light intensity is the light intensity at the *compensation depth*, where the photosynthesis and respiration of cells are in balance (Figure L2).

Formula

$$I_c = 0.2 I_0 e^{kD_c}$$

where

I_c = compensation light intensity ($\mu E/m^2 \cdot s$)
I_0 = incident light intensity ($\mu E/m^2 \cdot s$)
k = extinction coefficient (m^{-1})
D_c = compensation depth (m)

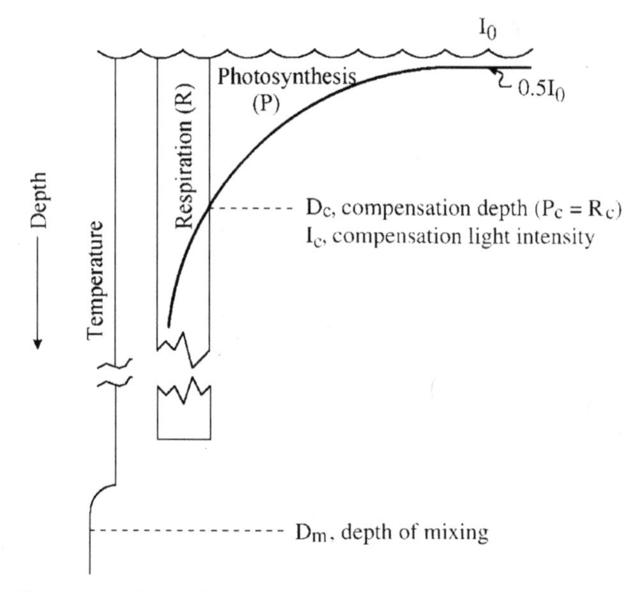

FIGURE L2: Compensation Light Intensity in Aquatic Environments (*Source:* Parsons and Takahashi, 1973.)

References

Parsons, T. R., and M. Takahashi, 1973. *Biological Oceanographic Processes*, Pergamon Press, Oxford.

Raymont, J. E. G. 1980. *Plankton and Productivity in the Oceans*, vol. 1, *Phytoplankton*, Pergamon Press, Oxford.

LIGHT: Relationship Between Light Frequency and Wavelength

Formula

Light frequency v is related to the light wavelength by the following formula:

$$v = \frac{c}{\lambda}$$

where

v = frequency (number of oscillations/s)
c = velocity of light = 3×10^{10} cm/s
λ = wavelength (cm)

Reference

Williams, W. R., W. L. Mattice, and H. B. Williams. 1978. *Basic Physical Chemistry for the Life Sciences*, W.H. Freeman, San Francisco.

LIGHT: Vertical Extinction Coefficient

See also LIGHT: Compensation Light Intensity.

Definition/Introduction

Light is rapidly absorbed as it penetrates water in a lake or other aquatic environments. Light intensity decreases exponentially with depth. The extinction (or attenuation) coefficient measures the loss of light with depth (Goldman and Horne, 1983). The vertical extinction coefficient is the slope of the line obtained by plotting the natural log of light intensity versus depth of the water (Figure L3).

Formula

The radiation intensity I_z at depth z is a function of incident light intensity I_0 and the extinction coefficient k according to the following formula:

$$I_z = I_0 e^{-kz}$$

where

I_z = light intensity (μE/m$^2 \cdot$ s) at depth z
I_0 = light intensity (μE/m$^2 \cdot$ s) immediately below the surface of the water
k = vertical extinction coefficient (m^{-1})
z = depth (m)

The vertical extinction coefficient k is given by the following equation (Schwoerbel, 1987):

$$k = \frac{\ln I_0 - \ln I_z}{z} = \frac{1}{z} \ln \frac{I_0}{I_z}$$

where $k = k_w + k_p + k_c$ (k_w, k_p, and k_c are the extinctions due the water itself, particulate matter, and color, respectively).

The terms *absorption coefficient* and *extinction coefficient* are used interchangeably when using natural logarithms (Cole, 1979). An approximate value of

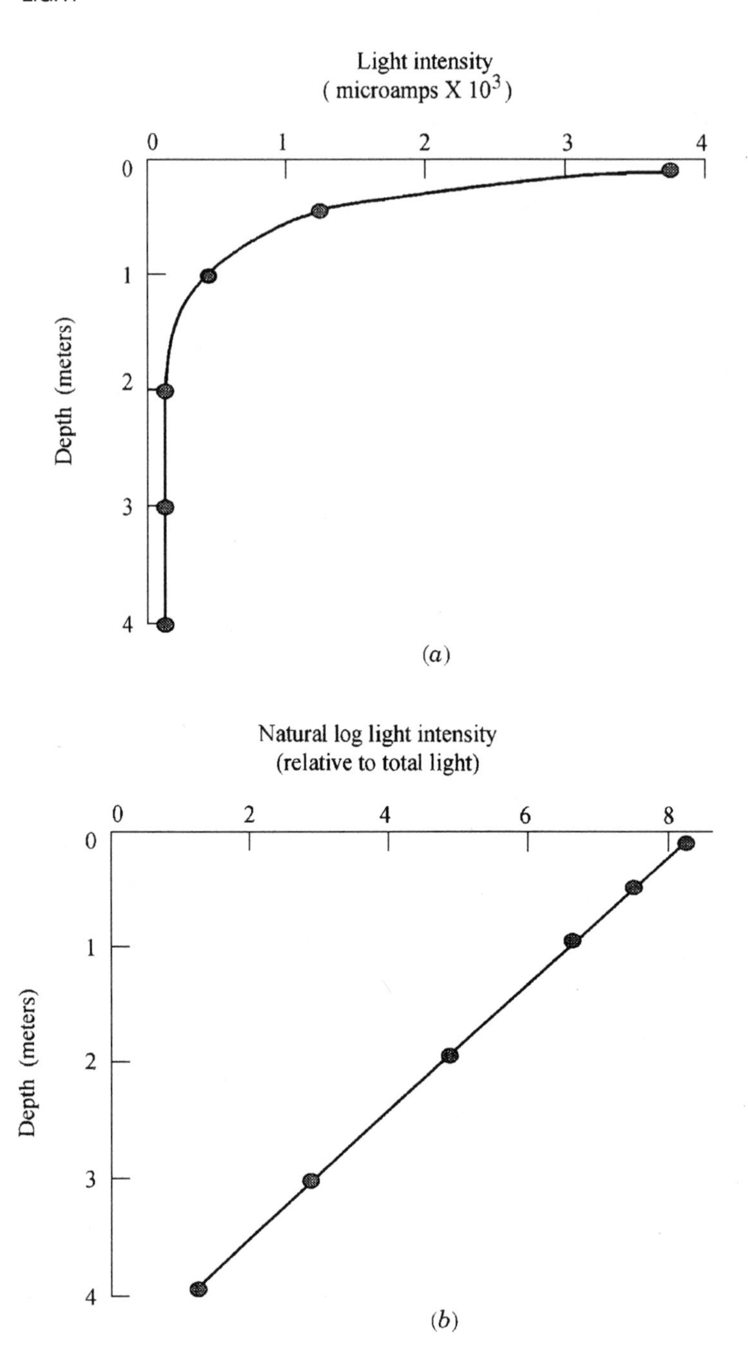

FIGURE L3: Light Intensity Versus Depth of the Water (*Sources:* Adapted from Lind, 1979.)

the extinction coefficient k can be obtained from Secchi disk readings (Raymont, 1980):

$$k = \frac{1.7}{z}$$

where z is the depth at the Secchi disk disappears from sight (m). For turbid waters, Holmes (1970) proposed the following relationship:

$$k = \frac{1.44}{z}$$

References

Cole, G. A. 1979. *Textbook of Limnology*, C.V. Mosby, St. Louis, MO.

Goldman, C. R., and A. J. Horne. 1983. *Limnology*, McGraw-Hill, New York.

Holms, R. W. 1970. The Secchi disk in turbid coastal waters. Limnol. Oceanogr. 15: 688–694.

Keller, A. 1989. Modeling the effects of temperature, light, and nutrients on primary productivity: an empirical and a mechanistic approach compared. Limnol. Oceanogr. 34: 82–95.

Lind, O. W. 1979. *Handbook of Common Methods in Limnology*, 2nd ed., C.V. Mosby, St. Louis, MO.

Raymont, J. E. G. 1980. *Plankton and Productivity in the Oceans*, vol. 1, *Phytoplankton*, 2nd ed., Pergamon Press, Oxford.

Schwoerbel, J. 1987. *Handbook of Limnology*, Ellis Horwood, Chichester, West Sussex, England.

LIGHT AVAILABILITY INDEX IN LAKES

Definition

In lakes, the light availability index, LI, expresses both the effects of the ratio of euphotic depth to mixing depth and the effect of day length.

Formula

$$\mathrm{LI} = 2\left(\frac{Z_s}{Z_m}\right)\frac{D}{24}$$

where

LI = light availability index

Z_s = Secchi disk depth (m) (euphotic depth = $2Z_s$)

Z_m = mixing depth (m)
D = daylength (h)

References

Makulla, A., and U. Sommer. 1993. Relationships between resource ratios and phytoplankton species composition during spring in five north German lakes. Limnol. Oceanogr. 38: 846–856.

Sommer, U. 1993. Phytoplankton competition in Plubsee: a field test of the resource ratio hypothesis. Limnol. Oceanogr. 38: 838–845.

LIGHT REFLECTION

Introduction

Light arriving at the surface of a body of water can be reflected, scattered, or absorbed by the water mass (for absorption, *see* LIGHT: Vertical Extinction Coefficient). A portion of the light reaching the surface of a water mass is immediately reflected. The amount of reflected light depends on the sun's angle of incidence, which varies with the latitude, the hour of the day, and the season (Cole, 1979).

Formulas

Light reflection or reflectance at the surface is given by Fresnel's formula (Schwoerbel, 1987):

$$R = \frac{1}{2}\left[\frac{\sin^2(i-r)}{\sin^2(i+r)} + \frac{\tan^2(i-r)}{\tan^2(i+r)}\right]$$

where

i = angle of incidence
r = angle of refraction

The following formula was proposed for quantifying light reflectance R (Anderson, 1952):

$$R = 1.18S^{-77}$$

where S is the sun's altitude or angular height in degrees. When the sun is at its zenith, the angle of incidence is zero, which means that there is almost no reflectance (Cole, 1979).

References

Anderson, E. R. 1952. Water-loss investigations. 1. Lake Hefner studies. Tech. Rep., U.S.G.S. Circ. 229: 71–119.

Cole, G. A. 1979. *Textbook of Limnology*, C.V. Mosby, St. Louis, MO.

Schwoerbel, J. 1987. *Handbook of Limnology*, Ellis Horwood, Chichester, West Sussex, England.

LINDEMAN'S EFFICIENCY

See Ecological Efficiencies.

LINEWEAVER–BURK EQUATION

Introduction

The Michaelis–Menten equation may be linearized by taking the reciprocal of both sides of the equation (*see* MICHAELIS–MENTEN EQUATION). One of the best known methods of linear transformation is referred to as the Lineweaver–Burk equation, which is represented by a *double-reciprocal* plot.

Formula

$$\frac{1}{V} = \frac{1}{V_{\max}} + \frac{K_m}{V_{\max}} \times \frac{1}{[S]}$$

where

V = reaction rate (units/time)

$V_{\max}$ = maximum reaction rate (units/time)

K_m = half saturation constant (Michaelis constant; the substrate concentration at which V is equal to $V_{\max}/2$)

[S] = substrate concentration (mol/L)

The plot of $1/V$ versus $1/[S]$ has a slope equal to $K_m/V_{\max}$, an intercept on the y-axis of $1/V_{\max}$ and an intercept on the x-axis of $-1/K_m$ (Figure L4). A disadvantage of the Lineweaver–Burk plot is the compression of the data at high substrate concentrations.

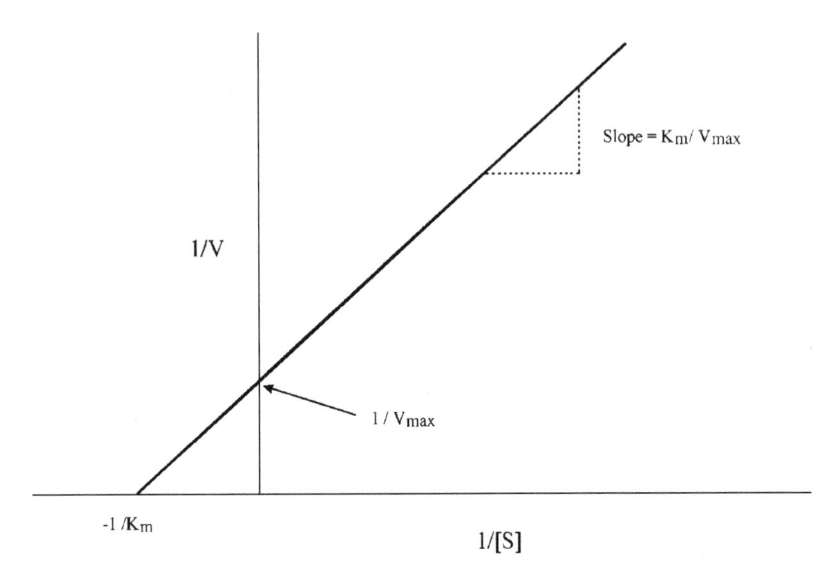

FIGURE L4: Typical Lineweaver–Burk Plot.

Another form of Lineweaver–Burk equation is obtained by multiplying both sides of the equation by [S] (Williams et al., 1978)

$$\frac{[S]}{V} = \frac{[S]}{V_{\max}} + \frac{K_m}{V_{\max}}$$

A plot of $[S]/V$ versus [S] gives a straight line with a slope of $1/V_{\max}$ and an intercept of $K_m/V_{\max}$ (see HANES EQUATION, Figure H2).

References

Dowd, J. E., and D. S. Riggs. 1965. A comparison of estimates of Michaelis–Menten kinetic constants from various linear transformations. J. Biol. Chem. 240: 863–869.

Segel, I. H. 1976. *Biochemical Calculations*, 2nd ed., Wiley, New York.

Williams, W. R., W. L. Mattice, and H. B. Williams. 1978. *Basic Physical Chemistry for the Life Sciences*, W.H. Freeman, San Francisco.

LIPOLYSACCHARIDE (LPS): Relation to Bacterial Biomass

Introduction

Lipopolysaccharide (LPS) is a component of the outer membrane of gram-negative bacteria. LPS determination is useful for the measurement of bacterial biomass in aquatic environments (fresh and marine waters), where most bacteria are gram negative. LPS is less useful in soils where the proportion of gram-positive bacteria is much higher.

Formula

Conversion of LPS to carbon content C (Watson et al., 1997):

$$C = 6.35 \text{ LPS}$$

Numerical Values

According to Watson et al. (1977):

- *E. coli* (laboratory conditions)
  Average LPS/cell = 39.40 fg
  Average carbon/cell = 207.5 fg
- *Marine bacteria* (average for 188 field samples)
  LPS/cell = 2.78 ± 1.42 fg

Reference

Watson, S. W., T. J. Novitsky, H. L. Quinby, and F. W. Valois. 1977. Determination of bacterial number and biomass in the marine environment. Appl. Environ. Microbiol. 33: 940–946.

LOTKA–VOLTERRA EQUATIONS

See also GROWTH, POPULATION.

Introduction

The Lotka–Volterra equations were proposed independently by Lotka and Volterra and describe the *competitive interaction between two species* using the same resource. They are extensions of the logistic equation of population growth (Lotka, 1925; Odum, 1971; Volterra, 1926).

Formulas

Species 1

$$\frac{dN_1}{dt} = r_{m1}N_1\frac{K_1 - N_1 - \alpha N_2}{K_1}$$

where

dN_1 = change in numbers of species 1

r_{m1} = maximum rate of increase of species 1

α = interspecific competition coefficient = inhibitory effect of species 2 on species 1
Also defined as the "conversion factor for expressing species 2 in units of species 1" ($N_1 - \alpha N_2$) (Krebs, 1972)

N_2 = number of individuals of species 2

K_1 = carrying capacity of the environment (or maximal value of N_1)

Species 2

$$\frac{dN_2}{dt} = r_{m2}N_2\frac{K_2 - N_2 - \beta N_1}{K_2}$$

where

DN_2 = change on numbers of species 2

r_{m2} = maximum rate of increase of species 2

N_2 = number of individuals of species 2

β = interspecific competition coefficient = inhibitory effect of species 1 on species 2

K_2 = carrying capacity of the environment (or maximal value of N_2)

The outcome of the competition between the two species will depend on the interspecific competition coefficients, α and β, and on the carrying capacities K_1 and K_2. There are four cases to consider:

$\alpha < K_1/K_2$; $\beta > K_2/K_1$: Only species 1 survives.
$\alpha > K_1/K_2$; $\beta < K_2/K_1$: Only species 2 survives.
$\alpha > K_1/K_2$; $\beta > K_2/K_1$: Equilibrium between the two species is unstable and either of the two species can survive.
$\alpha < K_1/K_2$; $\beta < K_2/K_1$: Both species survive.

References

Krebs, C. J. 1972. *Ecology*, Harper & Row, New York.

Lotka, A. J. 1925. *Elements of Physical Biology*, Williams & Wilkins, Baltimore.

McNaughton, S. J., and L. L. Wolf. 1979. *General Ecology*, 2nd ed., Holt, Rinehart and Winston, New York.

Odum, E. 1971. *Fundamentals of Ecology*, 3rd ed., W.B. Saunders, Philadelphia.

Smith, R. L. 1996. *Ecology and Field Biology*, 5th ed., Harper Collins, New York.

Volterra, V. 1926. Variazióne e flùttuazioni del nùmero d'indivìdui in spècie animali convivènti. Mem. Acad. Lincei 2: 31–113.

M

MAINTENANCE ENERGY OF CELLS

Definition

A growing microorganism uses a given substrate for growth (i.e., biomass production) and for maintaining the integrity of the cell. Thus the maintenance energy of cells m is the amount of energy that is used for purposes other than growth, which include motility, transport of nutrients into the cell against a concentration gradient or cell repair.

Formula

The rate of substrate utilization $dS/dt(= \mu X/Y_o)$ is equal to the rate of substrate utilization for biomass production ($\mu X/Y_g$) plus the rate of substrate utilization for maintenance (mX):

$$\frac{dS}{dt} = \frac{\mu X}{Y_o} = \frac{\mu X}{Y_g} + mX$$

or

$$\frac{1}{Y_o} = \frac{1}{Y_g} + \frac{m}{\mu}$$

where

S = limiting substrate concentration (mg/L)
μ = specific growth rate (day^{-1})
X = biomass (g dry weight)
Y_o = observed yield for substrate (g dry weight cells produced/g substrate used)
Y_g = true yield for growth (i.e., no energy is used for maintenance) (g dry weight cells produced/g substrate used)
m = specific maintenance coefficient (day^{-1})

If one plots $1/Y_o$ against $1/\mu$, one obtains a straight line with a slope of m and with $1/Y_g$ as the intercept on the abscissa.

Numerical Values

Numerical values for m: (m expressed in g substrate/g cell dry weight.day) (Anderson and Domsch, 1985; Morgan and Winstanley, 1996; Richard et al., 1985)

Filamentous bacterium type 021N (isolate N2) $\rightarrow m = 0.78$
Filamentous bacterium type 021N (isolate N7) $\rightarrow m = 0.32$

For bacteria in pure cultures the maintenance coefficient m is approximately 0.96 g glucose/g C biomass.day (Anderson and Domasch, 1985; Morgan and Winstanley, 1996). This is equivalent to 0.45 g glucose/g cell dry weight.day. We assume that cell C/cell dry weight = 0.45 (see DRY WEIGHT of microbial cells).

In soils: $m = 0.004$ g glucose/g C biomass.day at 28°C (= 0.00187 g glucose/g cell dry weight.day). At 15°C, $m = 0.00038$ g glucose/g C biomass.day (Anderson and Domsch, 1985).

References

Anderson, T. H., and K. H. Domsch. 1985. Determination of ecophysiological maintenance requirements of soil microorganisms in a dormant state. Biol. Fertil. Soils 1: 81–89.

Morgan, J. A. W., and C. Winstanley. 1996. Survival of bacteria in the environment, pp. 226–241, In: *Molecular Approaches to Environmental Microbiology*, R. W. Pickup and J. R. Sauders, Eds., Ellis Horwood, London.

Pirt, S. J. 1965. The maintenance energy of bacteria in growing cultures. Proc. R. Soc. B 163: 224–228.

Pirt, S. J. 1982. Maintenance energy: a general model for energy-limited and energy-sufficient growth. Arch. Microbiol. 133: 300–307.

Richard, M. G., G. P. Shimizu, and D. Jenkins. 1985. The growth physiology of the filamentous organism type 021N and its significance to activated sludge bulking. J. Water Pollut. Control Fed. 57: 1152–1162.

Slater, J. H. 1979. Microbial population and community dynamics, pp. 45–63. In: *Microbial Ecology: A Conceptual Approach*, J. M. Lynch and N. J. Poole, Eds., Blackwell Scientific, Oxford.

Stanier, R. Y., L. L. Ingraham, M. L., Wheelis, and P. R. Painter. 1986. *The Microbial World*, Prentice Hall, Upper Saddle River, NJ.

MARGALEF INDEX

See DIVERSITY INDEX.

MAXIMUM ACCEPTABLE TOXICANT CONCENTRATION (MATC)

Definition

The chronic toxicity of a chemical is expressed as maximum acceptable toxicant concentration. MATC is the threshold chemical concentration that produces a statistically significant effect on a given population (Rand and Petrocelli, 1985).

Formula

$$\text{NOEC} < \text{MATC} < \text{LOEC}$$

where

NOEC (no observed effect concentration)
= highest concentration that produces no effect on the test organisms) (mg/L)

LOEC (lowest observed effect concentration)
= lowest concentration of chemical that produces a statistically significant effect on the test organisms (mg/L)

Reference

Rand, G. M., and S. R. Petrocelli. 1985. *Fundamentals of Aquatic Toxicology*, Hemisphere, Washington, DC.

McINTOSH INDEX

See DIVERSITY INDEX.

MEAN CELL RESIDENCE TIME (MCRT) IN A BIOREACTOR

Definition

The mean cell residence time (MCRT) in a bioreactor is the mean residence time of microorganisms within the mixed bioreactor. It is also called the *solids retention time* (SRT) or *sludge age*.

Formula

$$\text{MCRT} = \frac{\text{mass of viable microorganisms in a reactor}}{\text{mass of viable microorganisms lost per unit time}}$$
$$= \frac{VX}{Q_w X + (Q - Q_w)X_e}$$

where

V = volume of aeration tank (m^3)
X = mixed liquor volatile suspended solids (mg/L)
Q_w = flow rate of wasted liquid (m^3/day)
Q = wastewater flow rate into the tank (m^3/day)
X_e = microorganism (suspended solids) concentration in effluent (mg/L)

References

Davis, M. L., and D. A. Cornwell. 1985. *Introduction to Environmental Engineering*, PWS, Boston.

Grady, C. P. L., and H. C. Lim. 1980. *Biological Wastewater Treatment: Theory and Applications*, Marcel Dekker, New York.

MEDIAN LETHAL CONCENTRATION (LC_{50})

Definition/Introduction

Toxicant dose–response relationship is one of the most basic concepts in toxicology. In safety evaluation of chemicals one must be able to measure the toxicity of a given chemical. Plotting the percent response (e.g., mortality) against the concentration of the test chemical gives a typical sigmoidal curve (Figure M1).

The median lethal concentration (LC_{50}) is the chemical concentration that produces mortality in 50% of the test population over a certain period of time. When effects other than mortality are used (e.g., behavioral or physiological effects), the term *median effective concentration* (EC_{50}) is used. The interpolation of LC_{50} or EC_{50} is shown in Figure M1.

Reference

Rand, G. M., and S. R. Petrocelli. 1985. *Fundamentals of Aquatic Toxicology*, Hemisphere, Washington, DC.

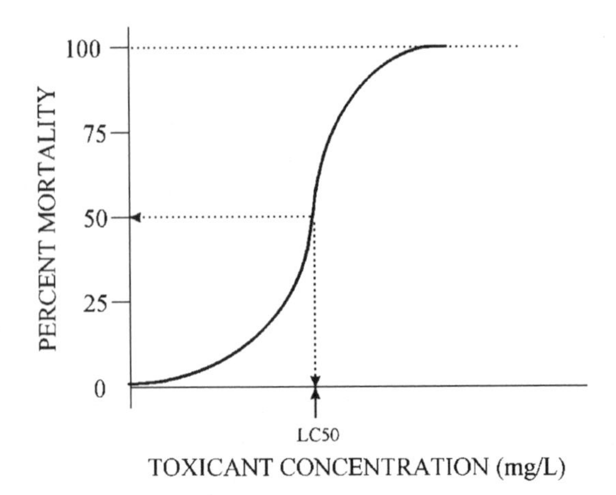

FIGURE M1: Median Lethal Concentration (LC_{50}) (*Source:* Rand and Petrocelli, 1985.)

MEMBRANE FILTRATION: Conversion of Number of Organisms per Field of View to Number of Organisms per Milliliter of Environmental Sample

Introduction

The microorganisms (e.g., bacteria, protozoa, yeast) present in an aquatic sample are stained, preferably using a fluorochrome (e.g., acridine orange, DAPI). The sample is passed through a membrane filter to retain the strained organisms. The filter is then observed via microscopy at a given magnification. Several (10–15) fields of view are examined, and the mean number of cells per field of view is derived.

Formula

The number of organisms per milliliter of sample is given by the following formula:

$$N = \frac{N_{\mathrm{fv}} S}{s(\mathrm{DF})V}$$

where

N = number of microorganisms/ml water

N_{fv} = cells/field of view (cm^2)

S = area of filter covered by sample (cm^2)

s = field of view area (cm^2) (determined using a stage micrometer)

DF = preservative dilution factor (ignore if no preservative is used)
V = mL of sample passed through the filter

Reference

Sherr, E. B., D. A. Caron, and B. F. Sherr. 1993. Staining of heterotrophic protists for visualization via epifluorescence microscopy, pp. 213–227. In: *Handbook of Methods in Aquatic Microbial Ecology*, P. F. Kemp, B. F. Sherr, E. B. Sherr, and J. J. Cole, Eds., Lewis Publishers, Boca Raton, FL.

METHANE: Conversion of BOD_5 of Solids in Facultative Ponds to Methane

Introduction

Biosolids (sludge) that settle out at the bottom of facultative ponds undergo anaerobic digestion, which produces methane and carbon dioxide.

Formula

The conversion of the BOD_5 of the solids to methane is given by the following relationship:

$$G = 6.28 \times 10^{-3}(T_s - 15)$$

where

G = solids (sludge) converted to gas (kg $BOD_5/m^2 \cdot day$)
T_s = sludge temperature (°C)

Reference

Rich, L. G. 1980. *Low-Maintenance, Mechanically Simple Wastewater Treatment Systems*, McGraw-Hill, New York.

METHANE PRODUCTION: Anaerobic Bioreactor

Introduction

The amount of methane produced in an anaerobic bioreactor depends on the removal of organic compounds (i.e., BOD removal) and the amount of volatile suspended solids (VSS) produced.

Formula

$$G = 0.351(S_r - 1.42X_v)$$

where

G = methane (CH_4) produced per day (m^3/day)

S_r = BOD removed per day (kg/day)

X_v = VSS removed per day (kg/day)

The heating value of 1 ft^3 (0.0283 m^3) of methane = 960 Btu = 1.01×10^6 J

Reference

Eckenfelder, W. W. Jr. 1989. *Industrial Water Pollution Control*, 2nd ed., McGraw-Hill, New York.

METHANE YIELD, THEORETICAL: Relationship with Ultimate Biochemical Oxygen Demand

Introduction

There is a relationship between methane production and the ultimate biochemical oxygen demand (BOD_L): 1 lb of BOD_L is equivalent to 0.25 lb of methane, which is equivalent to 5.62 ft^3 at 32°F and 1 atm pressure.

Reference

Metcalf and Eddy, Inc. 1991. *Wastewater Engineering: Treatment, Disposal and Reuse*, 3rd ed., McGraw-Hill, New York.

MICHAELIS–MENTEN EQUATION

Introduction

In 1913, Michaelis and Menten developed an expression for the kinetics of enzyme reactions. The expression applies to a reaction in which a substrate S combines with the active site of an enzyme molecules to form an enzyme–substrate complex (ES), which yields a new product (P), and the unchanged enzyme E is ready to react again with the substrate. At low substrate concentration, the reaction rate V is proportional to the substrate concentration (first-order kinetics). At higher substrate concentrations, V reaches a plateau

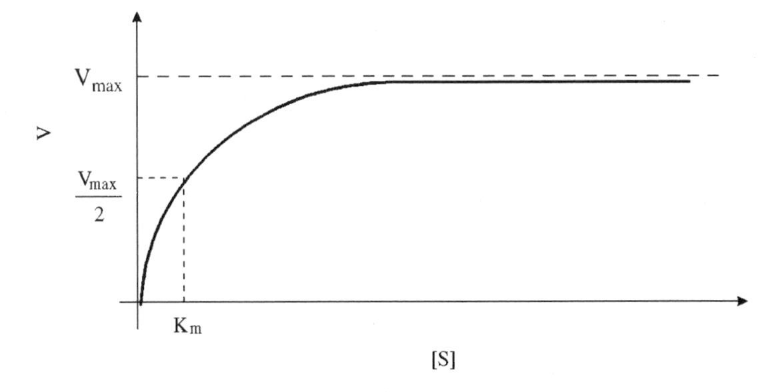

FIGURE M2: Michaelis–Menten Plot.

(zero-order kinetics). The enzymatic reaction rate V is given by the Michaelis–Menten equation (Figure M2).

Formula

$$V = \frac{V_{max}[S]}{K_m + [S]}$$

where

V = reaction rate (units/time)

V_{max} = maximum reaction rate (units/time)

[S] = substrate concentration (mol/L)

K_m = half saturation constant (Michaelis constant; the substrate concentration at which V is equal to $V_{max}/2$) (mol/L)

The parameters V_{max} and K_m can be determined from experimental data, using linearized or integrated versions of the Michaelis–Menten equation (*see* EADIE–SCATCHARD PLOT; HANES EQUATION; HOFSTEE EQUATION; LINEWEAVER–BURK EQUATION). Some of these linearization models give a biased estimate of the parameters, and the best approach for estimating these parameters is the nonlinear least squares analysis, which can be carried out with readily available software (Berthouex and Brown, 1994).

References

Berthouex, P. M., and L. C. Brown. 1994. *Statistics for Environmental Engineers*, Lewis Publishers, Boca Raton, FL.

Michaelis, L., and M. L. Menten. 1913. Biochem. Z. 49: 333–369.

MICROSCOPE: Resolving Power

Definition

The resolving power of a lens is the smallest distance two objects may be separated and still be distinguished from one another. The resolving power (RP) of a lens system is the size of the smallest object that can be seen through a given lens.

Formula

$$\text{Resolving power (RP)} = \frac{\lambda}{\text{NA}_{\text{obj}} + \text{NA}_{\text{cond}}}$$

where

λ = light wavelength (nm)

NA_{obj} = numerical aperture of objective

NA_{cond} = numerical aperture of condenser

Some give the resolving power as

$$\text{RP} = \frac{\lambda}{2 \times \text{NA}}$$

The numerical aperture of a lens system, which is given by the manufacturer, is given by the following:

$$\text{NA} = n \sin \theta$$

where

n = refractive index (for air, $n = 1.0$; for immersion oil, $n = 1.56$)

θ = half-angle of the cone of light entering the objective lens or exiting the condenser lens (see Figure M3).

References

Alcamo, I. E. 1983. *Fundamentals of Microbiology*, Addison-Wesley, Reading, MA.

Kerr, T. J. 1981. *Applications in General Microbiology: A Laboratory Manual*, Hunter's Textbooks, Winston-Salem, NC.

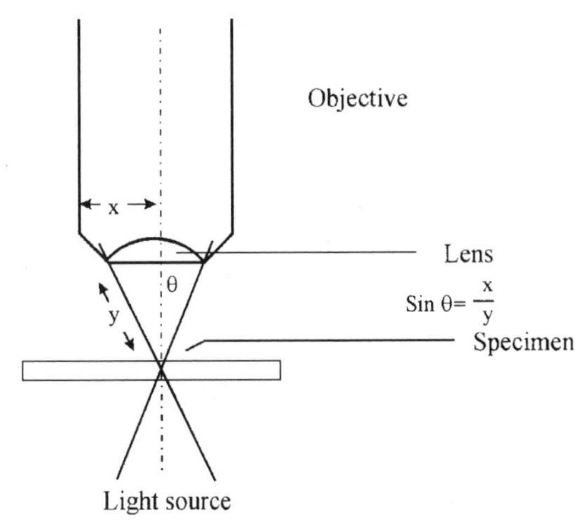

FIGURE M3: Determination of Numerical Aperture of a Microscope Lens (*Source:* Kerr, 1981.)

MIXED LIQUOR

Definition

Mixed liquor is the mixture of influent wastewater and return sludge in an activated sludge process.

MIXED LIQUOR SUSPENDED SOLIDS (MLSS)

Definition/Introduction

Mixed liquor suspended solids (MLSS) is the particulate solid concentration (measured as dry weight) in the mixed liquor. It is the total amount of organic and mineral suspended solids, including microorganisms, in the mixed liquor. MLSS is determined by filtering an aliquot of mixed liquor, drying the filter at 105°C and determining the weight of solids in the sample.

Numerical Values

See Table M1.

Reference

Metcalf and Eddy, Inc. 1979. *Wastewater Engineering: Treatment, Disposal and Reuse*, 2nd ed., McGraw-Hill, New York.

TABLE M1: Range of MLSS in Some Activated Sludge Processes

Process	MLSS (mg/L)
Conventional	1,500–3,000
Step aeration	2,000–3,500
Extended aeration	3,000–6,000
Pure oxygen system	6,000–8,000
High rate aeration	4,000–10,000

Source: Adapted from Metcalf and Eddy (1979).

MIXED LIQUOR VOLATILE SUSPENDED SOLIDS (MLVSS)

Definition

Mixed liquor volatile suspended solids (MLVSS) is the particulate solid concentration (measured as ash-free dry weight) in the mixed liquor. The organic portion of MLSS is represented by MLVSS, which comprises nonmicrobial organic matter as well as dead and live microorganisms, and cellular debris (Nelson and Lawrence, 1980). MLVSS is determined following heating of dried filtered samples at 600–650°C.

Numerical Values

Range The ratio MLVSS/MLSS in activated sludge ranges typically between 0.65 and 0.90.

Active Bacterial Content The proportion of active bacteria in the MLVSS ranges from <1% to as high as 50%, depending on the F/M (food-to-microorganism ratio) and the concentration of inert particulate organic matter in wastewater (Weddle and Jenkins, 1975).

References

Metcalf and Eddy, Inc. 1991. *Wastewater Engineering: Treatment, Disposal and Reuse*, 3rd ed. McGraw-Hill, New York.

Nelson, P. O., and A. W. Lawrence. 1980. Microbial viability measurements and activated sludge kinetics. Water Res. 14: 217–225.

Weddle, C., and D. Jenkins. 1975. The activity, viability, and bacterial content of activated sludge in relation to organic loading. Water Res. 8: 1322–1332.

MODELING: Mathematical Relations Commonly Encountered in Modeling

Introduction

Figure M4 shows the equations that describe four relationships (linear, exponential, logistic, Michaelis–Menten) commonly used in modeling.

Reference

Hall, C. A. S., and J. W. Day, Jr. 1990. Systems and models: terms and basic principles, In: *Ecosystem Modeling in Theory and Practice*, C. A. S. Hall and J. W. Day, Jr., Eds., University of Colorado Press, Boulder, CO.

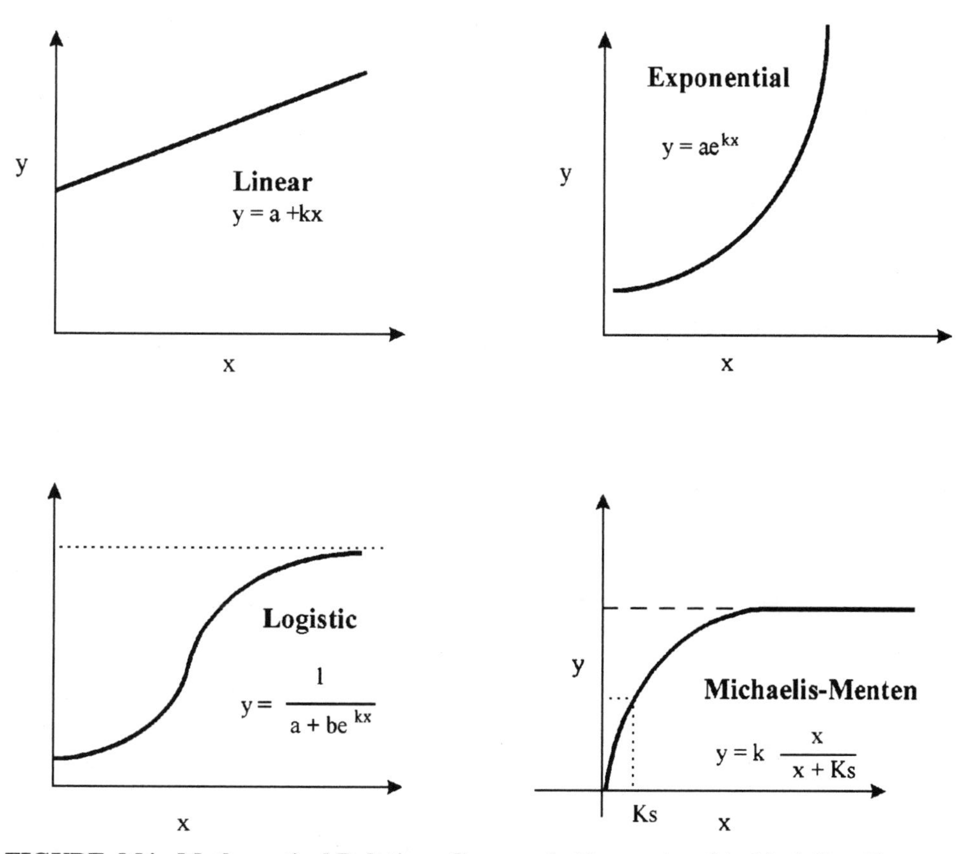

FIGURE M4: Mathematical Relations Commonly Encountered in Modeling (*Source:* Adapted from Hall and Day, Jr., 1990.)

MONOD'S EQUATION

Introduction

There are cases where growth is dependent of a single growth-limiting substrate with a concentration S. Monod's equation describes the relationship between microbial growth and the substrate concentration [S].

Formula

$$\mu = \mu_{\max} \frac{[\mathrm{S}]}{K_s + [\mathrm{S}]}$$

where

μ = specific growth rate (h^{-1})

$\mu_{\max}$ = maximum specific growth rate (h^{-1})

[S] = substrate concentration (mg/L)

K_s = half saturation constant (mg/L); is the substrate concentration at which the specific growth rate is equal to $\mu_{\max}/2$. K_s represents the affinity of the microorganism for the substrate (Figure M5)

There is a relationship between specific growth rate μ and cell residence time θ in a reactor:

$$\mu = \frac{1}{\theta} = Yq - k_d$$

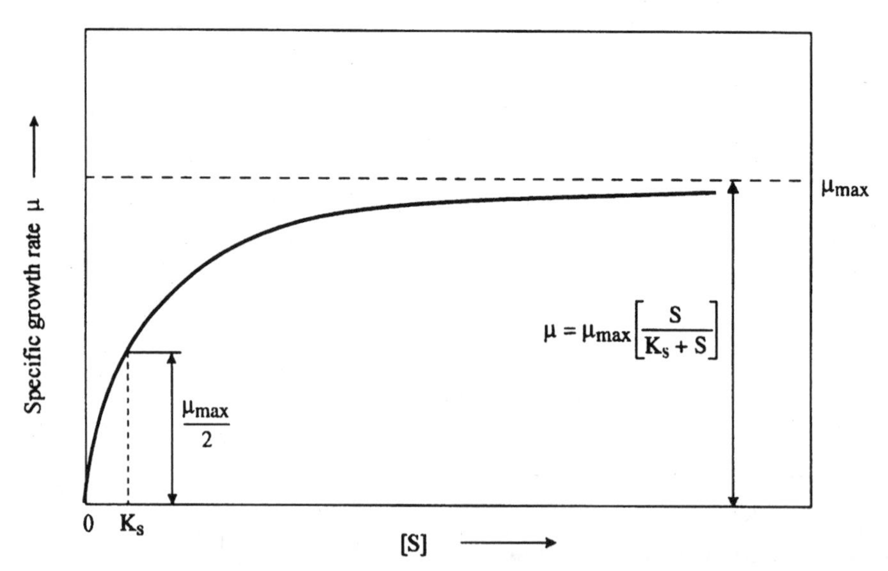

FIGURE M5: Relationship Between Microbial Specific Growth Rate and Substrate Concentration According to Monod's Equation.

where

Y = growth yield (g cells/g substrate utilized)

q = substrate utilization rate (g substrate utilized/g cells · h)

K_d = decay rate of microorganisms (h^{-1})

Numerical Values

The maximum specific growth rate of microorganisms μ_{max} depends on the microorganism type as well as on growth conditions (composition of the growth medium, temperature, pH etc.).

Bacteria See Table M2. Sikyta (1995) cites $\mu_{max} = 14.4–28.8 \text{ day}^{-1}$ for bacteria in general at 37°C.

For actinomycetes, the mean value reported was $\mu_{max} = 2.4–7.2 \text{ day}^{-1}$. For nitrifying bacteria in activated sludge, μ_{max} was equal to 0.4 (Horan, 1990).

Marine Bacteria The growth rates of bacteria from various marine environments was summarized and was found to vary from 0.007 to 6.9 day^{-1} (Moriarty, 1986).

Fungi See Table M3. Sikyta (1995) reported a general value for fungi: $\mu_{max} = 0.1–0.3 \text{ h}^{-1}$ at 28°C. For yeasts, $\mu_{max} = 0.3–0.5 \text{ h}^{-1}$ at 28°C.

Other Microorganisms See Table M4.

TABLE M2: Growth Kinetics at 15°C of Some Bacteria Isolated from Drinking Water

Substrate	Bacterium	K_s (μg C/L)	μ_{max} (h^{-1})
Glucose	*Flavobacterium* sp.	3.3	0.21
	Aeromonas hydrophila	16	0.28
	Pseudomonas fluorescens	57	0.22
	Enterobacter sp.	60	0.21
Glucose	*Flavobacterium* sp.	109	0.15
Acetate	*Pseudomonas fluorescens*	4	0.18
	Pseudomonas aeruginosa	28	0.09
Starch	*Flavobacterium* sp.	8.4	0.41

Source: Data summarized by van der Kooij (1995).

TABLE M3: μ_{max} and Doubling Times of Some Fungi Grown on Glucose

Fungus	μ_{max} (h^{-1})	Doubling Time (h)
Aspergillus niger (30°C)	0.20	3.4
Geotrichum candidum (25°C)	0.41	1.7
Penicillium chrysogenum (25°C)	0.12	5.6
Verticillium agaricinum (25°C)	0.24	2.9

Source: Adapted from Anderson et al. (1975).

TABLE M4: Maximum Growth Rates of Some Protists and Algae in Culture

Protist (Size in μm^3)	Temperature (°C)	μ_{max} (day^{-1})
Microflagellates		
Monas sp. (30)	20	1.2–4.8
Actinomonas mirabilis (75)	20	6.0
Ochromonas sp. (200)	20	4.6
Peuromonas jaculans (50)	20	3.8
Dinoflagellates		
Gymnodinium sp. (1000–1500)	1	0.3
Gymnodinium sp. (600–1200)	12	0.7
Heterocapsa pygmaea (600) (marine)	20	0.76
Ciliates		
Tintinnopsis acuminata (11,000)	18	1.4
Tintinnopsis vasculum (120,000)	5	0.5
Strombidium reticulatum (40,000)	12	0.9
Urotricha furcata (1800)	5.5	0.45
Uronema marinum (feeding on *Enterobacter aerogenes*)	21.5	1.72 3.86
Uronema nigricans (feeding on *enterobacter aerogenes*)		2.35
Marine phytoplankton		
Thalassiosira weissflogii	20	2.6
Thalassiosira oceanica	20	1.7
Dunalliela tertiolecta	20	1.9
Tetraselmis maculata	20	1.6

Source: Data from Sherr and Sherr (1994), Sherr et al. (1983), Ahner et al. (1995), Perez-Uz (1996).

References

Ahner, B. A., S. Kong, and F. M. M. Morel. 1995. Phytochelatin production in marine algae. 1. An interspecies comparison. Limnol. Oceanogr. 40: 649–657.

Anderson, C., J. Longton, C. Maddix, C. W. Scammell, and G. L. Solomons. 1975. The growth of microfungi on carbohydrates, pp. 314–329. In: *Single Cell Protein*, vol. 2, S. R. Tannenbaum and D. I. C. Wang, Eds., MIT Press, Cambridge, MA.

Horan, N. J. 1990. *Biological Wastewater Treatment Systems: Theory and Operation*, Wiley, New York.

Monod, J. 1942. Recherches sur la croissance des cultures bactériennes, 2nd ed., Hermann, Paris.

Monod, J. 1949. The growth of bacterial cultures. Annu. Rev. Microbiol. 3: 371–394.

Moriarty, D. J. W. 1986. Measurement of bacterial growth rates in aquatic systems from rates of nucleic acid synthesis. Adv. Microb. Ecol. 9: 245–292.

Perez-Uz, B. 1996. Bacterial preferences and growth kinetic variation *in Uronema marinum and Uronema nigricans* (Ciliophora: Scuticociliatida). Microb. Ecol. 31: 189–198.

Sherr, E. B., and B. F. Sherr. 1994. Bacterivory and herbivory: key roles of phagotrophic protists in pelagic food webs. Microb. Ecol. 28: 223–235.

Sherr, B. F., E. G. Sherr, and T. Berman. 1983. Grazing, growth, and ammonium excretion rates of a heterotrophic microflagellate fed with four species of bacteria. Appl. Environ. Microbiol. 45: 1196–1201.

Sikyta, B. 1995. *Techniques in Applied Microbiology*, Elsevier, Amsterdam.

van der Kooij, D. 1995. Significance and assessment of the biological stability of drinking water, pp. 89–102. In: *Water Pollution: Quality and Treatment of Drinking Water*, Springer–Verlag, New York.

MOST PROBABLE NUMBER (MPN)

Introduction

The most probable number (MPN) method helps estimate the number of organisms in a sample, using probability tables. Decimal dilutions of a given sample are incubated in a specific growth medium, and positive tubes (e.g., growth and gas production) are scored. MPN is given in tables or by using a formula.

Formulas

MPN can be computed using Thomas's formula, based on positive and negative tubes

$$\text{MPN}/100\ \text{mL} = \frac{\text{Number of positive tubes} \times 100}{\sqrt{\text{ml sample in negative tubes} \times \text{ml sample in all tubes}}}$$

MPN index is generally obtained from tables published in various books and handbooks. Tables M5, M6, and M7 give the MPN index/100 mL when using five 20-mL portions, ten 10-mL portions, or a series of three dilutions, with five tubes per dilution (10 mL, 1.0 mL, and 0.1 mL of sample), respectively (American Public Health Association, 1992).

If the series of decimal dilutions is different from that shown in the MPN tables, the following formula should be used to calculate the MPN index/100 mL:

$$\text{MPN index}/100\ \text{mL} = \frac{10 \times \text{MPN value from MPN table}}{\text{largest volume tested}}$$

References

American Public Health Association. 1992. *Standard Methods for the Examination of Water and Wastewater*, 18th ed., APHA, Washington, DC.

Bordner, R., and J. Winter. 1978. *Microbiological Methods for Monitoring the Environment*, EPA-600/8-78-017, U.S. Environmental Protection Agency, Cincinnati, OH.

Thomas, H. A., Jr. 1942. Bacterial densities from fermentation tubes. J. Am. Water Works Assoc. 34: 572–579.

TABLE M5: MPN Index and 95% Confience Limits for Various Combinations of Positive and Negative Results When Five 20-ml Portions are Used

No. of Tubes Giving Positive Reaction Out of 5 of 20 mL Each	MPN Index/100 mL	95% Confidence Limits (Approximate)	
		Lower	Upper
0	<1.1	0	3.0
1	1.1	0.05	6.3
2	2.6	0.3	9.6
3	4.6	0.8	14.7
4	8.0	1.7	26.4
5	>8.0	4.0	Infinite

Source: Standard Methods for the Examination of Water and Wastewater, 1992 (with permission)

TABLE M6: MPN Index and 95% Confience Limits for Various Combinations of Positive and Negative Results When Ten 10-ml Portions are Used

No. of Tubes Giving Positive Reaction Out of 10 of 10 mL Each	MPN Index/100 mL	95% Confidence Limits (Approximate)	
		Lower	Upper
0	<1.1	0	3.0
1	1.1	0.03	5.9
2	2.2	0.26	8.1
3	3.6	0.69	10.6
4	5.1	1.3	13.4
5	6.9	2.1	16.8
6	9.2	3.1	21.1
7	12.0	4.3	27.1
8	16.1	5.9	36.8
9	23.0	8.1	59.5
10	23.0	13.5	Infinite

Source: Standard Methods for the Examination of Water and Wastewater, 1992 (with permission)

MURAMIC ACID

See also BIOMASS BACTERIAL.

Introduction

Muramic acid is a component of the cell wall of prokaryotic microorganisms (bacteria and blue-green algae). It is determined by converting muramic acid to lactate which is then determined by enzymatic (Moriarty, 1977) or chemical analysis (King and White, 1977).

Muramic Acid Content of Microorganisms Gram-positive bacteria have higher muramic acid content than gram-negative bacteria. Thus one must estimate the proportion of these two groups of bacteria. (*Note:* Muramic acid is given in μg/mg C or in μg/mg cell dry weight. To convert from one to the other, we assume that C is about 45% of cell dry weight.)

Numerical Values

- *Soil bacteria grown under laboratory conditions* (Millar and Casida, 1970; King and White, 1977)

TABLE M7: MPN Index and 95% Confidence Limits for Various Combinations of Positive Results When Five Tubes are Used per Dilution (10 ml, 1.0 ml, 0.1 ml) Are Used

Combination of Positives	MPN Index/ 100 mL	95% Confidence Limits		Combination of Positives	MPN Index/ 100 mL	95% Confidence Limits	
		Lower	Upper			Lower	Upper
				4-2-0	22	9.0	56
0-0-0	<2	—	—	4-2-1	26	12	65
0-0-1	2	1.0	10	4-3-0	27	12	67
0-1-0	2	1.0	10	4-3-1	33	15	77
0-2-0	4	1.0	13	4-4-0	34	16	80
				5-0-0	23	9.0	86
1-0-0	2	1.0	11	5-0-1	30	10	110
1-0-1	4	1.0	15	5-0-2	40	20	140
1-1-0	4	1.0	15	5-1-0	30	10	120
1-1-1	6	2.0	18	5-1-1	50	20	150
1-2-0	6	2.0	18	5-1-2	60	30	180
2-0-0	4	1.0	17	5-2-0	50	20	170
2-0-1	7	2.0	20	5-2-1	70	30	210
2-1-0	7	2.0	21	5-2-2	90	40	250
2-1-1	9	3.0	24	5-3-0	80	30	250
2-2-0	9	3.0	25	5-3-1	110	40	300
2-3-0	12	5.0	29	5-3-2	140	60	360
3-0-0	8	3.0	24	5-3-3	170	80	410
3-0-1	11	4.0	29	5-4-0	130	50	390
3-1-0	11	4.0	29	5-4-1	170	70	480
3-1-1	14	6.0	35	5-4-2	220	100	580
3-2-0	14	6.0	35	5-4-3	280	120	690
3-2-1	17	7.0	40	5-4-4	350	160	820
				5-5-0	240	100	940
4-0-0	13	5.0	38	5-5-1	300	100	1300
4-0-1	17	7.0	45	5-5-2	500	200	2000
4-1-0	17	7.0	46	5-5-3	900	300	2900
4-1-1	21	9.0	55	5-5-4	1600	600	5300
4-1-2	26	12	63	5-5-5	⩾ 1600	—	—

Source: Standard Methods for the Examination of Water and Wastewater, 1992 (with permission)

Gram-negative bacteria: 3.44 ± 0.5 μg/mg dry weight
Gram-positive bacteria: 9.6 ± 1.90 μg/mg dry weight
Spores of gram-positive bacteria: 38 ± 6.2 μg/mg dry weight

- *Sediment bacteria* (Moriarty, 1975, 1977)
  Gram-negative rods: 20 μg/mg C
- *Cultured bacterial cells* (Moriarty, 1977)
  Gram-negative bacteria: 7.5–12 μg/mg C
  Gram-positive bacteria: 44 μg/mg C
  Blue-green algae (*Oscillatoria tenuis*): 11 μg/mg C
- *Marine sediments* (Moriarty, 1977): To take into account the proportions of gram-positive and gram-negative bacteria, the following relationship was proposed:

$$C = \frac{1000M}{12n + 40p}$$

where

C = bacterial biomass carbon (mg)
M = muramic acid content (μg)
n = proportion of gram-negative and weakly gram-positive bacteria
p = proportion of gram-positive bacteria

Others suggested a conversion factor of 6.4–12 μg muramic acid/mg bacterial carbon (Jones, 1979). The estimation of biomass from muramic acid is difficult in natural environments, due to the presence of mixed microbial populations (Herbert, 1990).

References

Herbert, R. A. 1990. Methods for enumerating microorganisms and determining biomass in natural environments, pp. 1–39. In: *Methods in Microbiology*, vol. 22, R. Grigorova and J. R. Norris, Eds., Academic Press, London.

Jones, J. G. 1979. *A Guide to Methods for Estimating Microbial Numbers and Biomass in Fresh Water*, Freshwater Biological Association, Windermere, Westmorland, England.

King, J. D., and D. C. White. 1977. Muramic acid as a measure of microbial biomass in estuarine and marine samples. Appl. Environ. Microbiol. 33: 777–783.

Millar, W. N., and L. E. Casida. 1970. Evidence for muramic acid in the soil. Can. J. Microbiol. 18: 299–304.

Moriarty, D. J. W. 1975. A method for estimating the biomass of bacteria in aquatic sediments and its application to trophic studies. Oecologia 20: 219–227.

Moriarty, D. J. W. 1977. Improved method using muramic acid to estimate biomass of bacteria in sediments. Oecologia 26: 317–323.

N

NERNST EQUATION

Introduction

The Nernst equation describes the flow of ions across cell membrane channel protein. The ion flow is driven by the electrochemical gradient, which is the result of the voltage gradient and the concentration gradient across the cell membrane. Equilibrium is reached at the equilibrium potential V of the ion, which is given by the Nernst equation.

Formulas

$$V = \frac{RT}{zF} \ln \frac{C_o}{C_i}$$

where

V = equilibrium potential (V)
R = gas constant
T = absolute temperature (K)
z = ion valence (ion charge)
F = Faraday's constant ($2.3\ 10^4$ cal/V · mol)
C_o = outside concentration of the ion
C_i = inside concentration of the ion

The relationship among electrode potential, temperature, and species concentrations for a given system is also expressed by another form of the Nernst equation (Weber, 1972):

$$E_h = E_0 + \frac{2.3RT}{nF} \log_{10} \frac{[\text{oxidant}]}{[\text{reductant}]}$$

where

E_0 = standard half reaction potential (V)
R = gas constant = 8.314 J/K · mol
T = absolute temperature (K)

[oxidant] = molar concentration of oxidant
[reductant] = molar concentration of reductant
n = number of electrons transferred
F = Faraday constant = 96,493 C

At 25°C, the equation above becomes (Weber, 1972)

$$E_h = E_0 + \frac{0.059}{n} \log_{10} \frac{[\text{oxidant}]}{[\text{reductant}]}$$

E_h is related to $p\varepsilon$ by the following:

$$E_h = 0.059 p\varepsilon$$

where $p\varepsilon$ is the hypothetical electron activity in solution and represents the relative tendency of a solution to accept electrons. Reducing solutions have low $p\varepsilon$, while oxidizing solutions have high $p\varepsilon$ (McBride, 1994).

Numerical Values

See Table N1.

References

Alberts, B., D. Bray, J. Lewis, M. Raff, K. Roberts and J. D. Watson. 1989. *Molecular Biology of the Cell*, 2nd ed., Garland, New York.

McBride, M. B. 1994. *Environmental Chemistry of Soils*, Oxford University Press, New York.

Weber, W. J. Jr., Ed. 1972. *Physicochemical Processes for Water Quality Control*, Wiley-Interscience, New York.

TABLE N1: E_0 for Some Chemical Reactions

Reaction	E_0 (V)
$2H^+ + 2e^- = H_2$	0.00
$Cu^{2+} + e^- = Cu^+$	0.153
$Fe^{3+} + e^- = Fe^{2+}$	0.77
$O_2 + 4H^+ + 4e^- = 2H_2O$	1.229
$Mn^{3+} + e^- = Mn^{2+}$	1.51

Source: Adapted from McBride (1994).

NITRIFICATION KINETICS: Effect of pH and Temperature on Growth Rate of nitrifiers

Introduction

Several formulas have been proposed for estimating the maximum growth rate ($\mu_{\max}$) of nitrifiers as a function of temperature and pH, which are major factors controlling nitrification.

Formula

Antoniou et al. (1990) proposed the following formula expressing the maximum growth rate of *Nitrosomonas* as a function of temperature and pH:

$$\mu_A b_A = \frac{me^{-(a/T)}}{1 + \frac{b}{10^{-\mathrm{pH}}} + \frac{10^{-\mathrm{pH}}}{c}}$$

where

μ_A = *Nitrosomonas* maximum specific growth rate (day^{-1})
b_A = *Nitrosomonas* decay rate (day^{-1})

and where m, a, b and c are coefficients that can be determined from experimental trials at different pH and temperature. The authors determined the following coefficient values using sludge from a modified Ludzack–Ettinger process in Gainesville, Florida:

$$m = 4.70 \times 10^{14}$$
$$a = 9.98 \times 10^{3}$$
$$b = 2.05 \times 10^{-9}$$
$$c = 1.66 \times 10^{-7}$$

Reference

Antoniou, P., J. Hamilton, B. Koopman, R. Jain, B. Holloway, G. Lyberatos, and S. A. Svoronos. 1990. Effect of temperature and pH on the effective maximum specific growth rate of nitrifying bacteria. Water Res. 24: 97–101.

NITROGEN/PHOSPHORUS MASS RATIOS

Introduction

The N/P ratios vary in aquatic environments. Oligotrophic lakes generally have higher N/P ratios than mesotrophic or eutrophic lakes. This ratio reflects the source of nutrients to lakes. Oligotrophic lakes receive inputs from natural sources (N/P ratios > 20), whereas eutrophic lakes receive inputs from anthropogenic sources such as wastewater (N/P ratios < 10) (Table N2).

TABLE N2: Average N/P Mass Ratios in Potential Nutrient Sources of Freshwater Lakes

Source	N/P
Runoff from unfertilized fields	247.4
Export from soils, medium fertility	75.0
Export from forested areas	71.1
Export from soils, fertile	33.3
Groundwater	28.5
Precipitation	25.4
Runoff, tropical forest	23.5
Precipitation	23.2
Export from agriculture watersheds	20.0
River water	18.9
River water (Mississippi)	12.2
Sewage	10.0
Seepage from cattle manure	8.9
Zooplankton excreta	8.9
Fertilizer, average	7.9
Precipitation, tropical	7.7
Feedlot runoff	6.4
Sediments, mesotrophic lake	6.3
Urban stormwater drainage	5.8
Sewage	5.3
Zooplankton excreta	5.0
Urban runoff	4.7
Sediments, oligotrophic lakes	3.3
Sewage	2.8
Septic tank effluent	2.7
Gull feces	0.8
Rocks, sedimentary	0.8
Rocks	< 0.1

Source: Adapted from Downing and McCauley (1992)

TABLE N3: Average N:P Mass Ratios in Aquatic Organisms

Component	N/P
Frog	17.8
Cyanobacteria, in blooms	16.3
Seston, hypertrophic	14.7
Fish, freshwater average	13.9
Macrophytes	
Oligotrophic lake	13.6
Tropical	18.3
Marine	8.3
Eutrophic lake	6.8
Watercress	6.5
Periphyton, hypertrophic	12.5
Algae	10.0
Planktonic arthropods, marine	9.2
Zooplankton	
Cladocera	8.8
Copepods	7.8
Algae, whole community	7.0
Daphnia spp.	6.6
Seston	6.4
Fish, whole	3.9
Bacteria (*Escherichia coli*)	3.2

Source: Adapted from Downing and McCauley (1992).

Numerical Values

Table N3 shows that the N/P ratios of aquatic organisms vary between 3.2 in *Escherichia coli* to 18.2 for tropical macrophytes.

Reference

Downing, J. A., and E. McCauley. 1992. The nitrogen:phosphorus relationships in lakes. Limnol. Oceanogr. 37: 936–945.

NITROGENEOUS BIOCHEMICAL OXYGEN DEMAND

Definition

Autotrophic bacteria such as nitrifying bacteria require oxygen to oxidize NH_4^+ to NO_3. The oxygen demand exerted by nitrifiers is called autotrophic BOD or nitrogeneous biochemical oxygen demand (NBOD).

Formula

$$NBOD = BOD - CBOD$$

where

NBOD = nitrogeneous oxygen demand (mg/L)
BOD = biochemical oxygen demand (mg/L)
CBOD = carbonaceous BOD (mg/L)

The theoretical nitrogeneous oxygen demand is 4.57 g oxygen used per gram of ammonium oxidized to nitrate. However, this value is actually lower and must be corrected due to incorporation of some of the nitrogen into the microbial cells (Davis and Cornwell, 1985). Thus NBOD is as follows (Verstraete and van Vaerenbergh, 1986):

$$NBOD(mg/L) = (\text{available N} - \text{assimilated N}) \times 4.33$$

It is necessary to carry out an inhibited BOD test to distinguish between carbonaceous (CBOD) and nitrogeneous BOD (NBOD) (American Public Health Association, 1989). It is recommended to add 2-chloro-6(trichloromethyl)pyridine at a final concentration of 10 mg/L for nitrification inhibition.

References

American Public Health Association. 1989. *Standard Methods for the Examination of Water and Wastewater*, 17th ed., APHA, Washington, DC.

Davis, M. L., and D. A. Cornwell. 1985. *Introduction to Environmental Engineering*, PWS, Boston.

Liu, C. C. K. 1986. Surface water quality analysis. In: *Handbook of Environmental Engineering*, vol. 4, *Water Resources and Natural Control Processes*, L. K. Wang and N. C. Pereira, Eds., Humana Press, Clifton, NJ.

Verstraete, W., and E. van Vaerenbergh. 1986. Aerobic activated sludge, pp. 43–112. In: *Biotechnology*, vol. 8, *Microbiol Degradations*, W. Schonborn, Ed., VCH, Weinheim, Germany.

NUTRIENT REQUIREMENTS: Suspended Growth Systems

Introduction

The empirical formula, $C_{60}H_{87}O_{23}N_{12}P$, may be used to approximate the elemental composition of bacterial cells. According to this formula, the nitrogen and phosphorus of bacterial cells is 12.2% and 2.3%, respectively. In bioreactors, the nitrogen and phosphorus requirements per unit of BOD_5 consumed decrease as the mean solids retention time increases.

Formula

$$BOD_5/N/P = \frac{R_s}{0.023R_x} : 5.3 : 1$$

where

BOD_5 = 5-day biochemical oxygen demand (mg/L)
R_s = rate of BOD_5 utilization (g/day)
R_x = net rate of biomass production (g/day)

The formulas for R_s and R_x are given below:

$$R_x = \frac{QY(S_i - S_e)}{1 + k_d q_s}$$

$$R_s = Q(S_i - S_e)$$

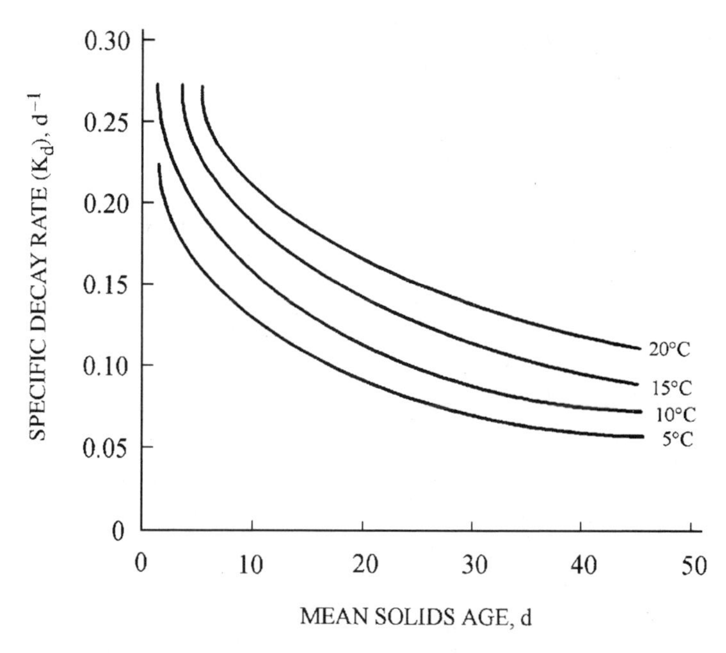

FIGURE N1: Specific Decay Rate as a Function of Temperature and Mean Solids Retention Time (*Source:* Adapted from Rich, 1980.)

where

Q = influent wastewater flow rate (m^3/day)
Y = true growth yield coefficient (g biomass produced/g BOD_5 removed)
S_i = influent BOD_5 concentration (mg/L)
S_e = effluent BOD_5 concentration (mg/L)
k_d = specific decay coefficient (day^{-1})
q_s = mean cell residence time (day)

q_s is given by the following equation:

$$\frac{1}{q_s} = \mu - k_d$$

where μ is the specific growth rate (day^{-1}).

Numerical values

A typical value of Y is 0.5 g VSS/g BOD_5. Values for k_d are given as a function of temperature and mean cell residence time in Figure N1.

Reference

Rich, L. G. 1980. *Low-Maintenance, Mechanically Simple Wastewater Treatment Systems*, McGraw-Hill, New York.

O

OCTANOL–WATER PARTITION COEFFICIENT (K_{ow})

See PARTITION COEFFICIENT: Octanol–Water.

ORGANIC LOADING RATE

Definition

In wastewater treatment systems, the rate of introduction or organic compounds is defined as the organic loading rate.

Formula

$$\mathrm{OLR} = \frac{\mathrm{BOD}}{\mathrm{MLSS}}\left(\frac{1}{t_r}\right)$$

where

- OLR = organic loading rate (kg BOD/kg MLSS · day)
- BOD = biochemical oxygen demand (mg/L)
- MLSS = mixed liquor suspended solids (mg/L)
- t_r = hydraulic detention time (day)

Numerical Value

In conventional activated sludge OLR is in the range 0.25–0.45 (Forster and Johnston, 1987).

Reference

Forster, C. F., and D. W. M. Johnston. 1987. Aerobic processes, pp. 15–56. In: *Environmental Biotechnology*, C. F. Forster and D. A. J. Wase, Eds., Ellis Horwood, Chichester, West Sussex, England.

OSMOTIC PRESSURE

Definition

Osmotic pressure is the difference in pressure between two solutions at equilibrium of varying salinities which are separated by a semipermeable membrane. It is a measure of the potential energy difference of the water molecules between the two solutions.

Formulas

$$\pi = \frac{RT}{V_A} \ln \frac{P_A^0}{P_A}$$

where

π = osmotic pressure (atm)
R = 0.082 (L · atm/(mol · K)
T = temperature (K)
V_A = volume/mole of solvent (L/mol) ($V_A = 0.018$ l for water)
P_A^0 = vapor pressure of solvent in the dilute solution (atm)
P_A = vapor pressure of solvent in the concentrated solution (atm)

For dilute solutions, the osmotic pressure is given by (Tinoco et al., 1995)

$$\pi = cRT$$

where c is the concentration of solute (mol/L).

References

Sawyer, C. N., and P. L. McCarty. 1978. *Chemistry for Environmental Engineering*, McGraw-Hill, New York.

Sundstrom, D. W., and H. E. Klei. 1979. *Wastewater Treatment*, Prentice Hall, Upper Saddle River, NJ.

Tinoco, I. Jr., K. Sauer, and J. C. Wang. 1995. *Physical Chemistry: Principles and Applications in Biological Sciences*, Prentice Hall, Upper Saddle River, NJ.

OXIDATION–REDUCTION POTENTIAL

See REDOX POTENTIAL.

TABLE O1: Critical Oxygen Concentrations, C_{crit}, of Some Microorganisms

Microorganisms	C_{crit} (mg/L)
Escheerichia coli	0.26
Pseudomonas ovalis	1.10
Penicillium chrysogenum	0.40
Saccharomyces cerevisiae	0.60
Torulopsis utilis	2.00

Source: Data from Brown (1970), Crueger and Crueger (1989).

OXYGEN CONCENTRATION, CRITICAL

Definition/Introduction

The critical oxygen concentration is the level at which microbial respiration occurs without any hindrance. At oxygen levels lower than the critical value, microbial respiration is correlated with the dissolved oxygen concentration. Above the critical value, there is no relationship between respiration rate and dissolved oxygen.

Numerical Values

See Table O1.

References

Brown, D. E. 1970. Aeration in the submerged culture of microorganisms, pp. 125–174. In: *Methods in Microbiology*, vol. 2, J. R. Norris and D. W. Ribbons, Eds., Academic Press, San Diego, CA.

Crueger, W., and A. Crueger. 1989. *Biotechnology: A Textbook of Industrial Microbiology*, Sinauer Associates, Sunderland, MA.

OXYGEN CONSUMPTION RATE

See ACTIVATED SLUDGE: Oxygen Consumption Rate.

OXYGEN DEFICIT IN A STREAM

Introduction

In a stream, oxygen deficit can occur as a result of discharge of a wastewater effluent into the stream.

Formula

$$D = C_s - C$$

where

D = oxygen deficit (mg/L)

C_s = saturation level of dissolved oxygen in the stream (mg/L) (depends on water temperature and on the amount of dissolved solids)

C = actual dissolved oxygen level in the stream (mg/L)

References

Metcalf and Eddy, Inc. 1991. *Wastewater Engineering: Treatment, Disposal and Reuse*, 3rd ed., McGraw-Hill, New York.

Ray, B. T. 1995. *Environmental Engineering*, PWS, Boston.

OXYGEN EQUIVALENTS

Introduction

Oxygen equivalents are also known as *theoretical oxygen demand* (TOD) and are expressed as g O_2/g substrate used.

Examples

- *Oxygen equivalents for glucose.* These can be determined from the following respiration equation:

$$C_6H_{12}O_6 + 6O_2 \rightarrow 6CO_2 + 6H_2O$$

  Oxygen equivalents $= 6 \times 32/180 = 1.06$ g O_2/g substrate.

- *Oxygen equivalents for nitrification.* These can be determined from the following nitrification equation:

$$NH_4^+ + 2O_2 \rightarrow NO_3^- + H_2O + 2H^+$$

Oxygen equivalents $= 2 \times 32/14 = 4.57\ g\ O_2/g\ NH_4{}^+$-N. This is called the *nitrogeneous oxygen demand.*

- *Oxygen equivalents of methane*

$$CH_4 + 2O_2 \rightarrow CO_2 + 2H_2O$$

Oxygen equivalents $= 2 \times 32/16 = 4\ g\ O_2/g\ CH_4$.

Reference

Andrews, J. F. 1983. Kinetics and mathematical modeling, pp. 113–172. In: *Ecological Aspects of Used-Water Treatment*, vol. 3, C. R. Curds and H. A. Hawkes, Eds., Academic Press, London.

OXYGEN PRODUCTION: Photosynthesis in Aerobic Ponds

Formula

The oxygen produced by photosynthesis in an aerobic pond can be estimated, using the following formula:

$$O_2 = CfS$$

where

$C = 2.8 \times 10^{-5}\ kg/m^2 \cdot day)$

f = light conversion efficiency (%)

S = light intensity ($cal/cm^2 \cdot day$) (varies with latitude and month of the year)

Reference

Eckenfelder, W. W. Jr., 1989. *Industrial Water Pollution Control*, 2nd ed., McGraw-Hill, New York.

OXYGEN REQUIREMENT: Suspended Growth Systems

Introduction

Under aerobic conditions, a portion of the organic substrate is oxidized to provide energy for growth and cell maintenance, while the remainder is utilized as a carbon source for cell synthesis. Additional oxygen is utilized in the oxidation of ammonia to nitrate.

Formula

The following equation is applicable under aerobic conditions

$$R_o = aQ(S_i - S_e) + bQ(N_i - N_e) - \frac{cVX_a}{q_s}$$

where

R_o = oxygen utilization rate (g O_2/day)
Q = volumetric flow rate in the system (m^3/day)
S_i = influent organic substrate concentration (g/m^3)
S_e = effluent organic substrate concentrations (g/m^3)
N_i, N_e = influent and effluent concentrations of ammonia nitrogen (g/m^3)
a, b, c = oxygen equivalent ratios
V = volume of aeration basin (m^3)
X_a = total bacterial cell concentration as VSS (g/m^3)
q_s = mean cell residence time (day)

The ratio a is assumed to have a value of 1.00 if the organic substrate is measured in terms of BOD_L (i.e., ultimate BOD) or COD or a value of 1.47 if the organic substrate is expressed in terms of 5-day BOD. The ratio b has a value of 4.57. The ratio c is assumed to be 1.42.

Numerical Values

See Table O2.

TABLE O2: Specific Oxygen Requirements of Pure Cultures of Microorganisms

Microorganisms	Oxygen Requirement (mM O_2/g cells · h)
Aspergillus niger	3.0
Pencicillium chrysogenum	3.9
Klebsiella aerogenes	4.0
Saccahromyces cerevisiae	8.0
Escherichia coli	10.0

Source: Adapted from Brown (1970).

References

Brown, D. E. 1970. Aeration in the submerged culture of microorganisms, pp. 125–174. In: *Methods in Microbiology*, vol. 2, J. R. Norris and D. W. Ribbons, Eds., Academic Press, San Diego, CA.

Rich, L. G. 1980. *Low-Maintenance, Mechanically Simple Wastewater Treatment Systems*, McGraw-Hill, New York.

OXYGEN TRANSFER IN TRICKLING FILTERS

Introduction

As air passes through a trickling filter, it is transferred to the liquid flowing over the biofilm that develops over the filter material. The rate of oxygen transfer is influenced by fluid mixing and turbulence and by the hydraulic loading (Eckenfelder, 1989).

Formula

$$N = 0.06K_0(C_s - C_L)Q$$

where

N = oxygen transfer ($kg\,O_2/m^3 \cdot h$)

K_0 = transfer-rate coefficient (m^{-1}). (K_0 decreases as the hydraulic loading Q increases)

C_s = oxygen saturation (mg/L)

C_L = dissolved oxygen concentration of the liquid pressing over the filter (mg/L)

Q = hydraulic loading ($m^3/m^2 \cdot min$)

Reference

Eckenfelder, W. W. 1989. *Industrial Water Pollution Control*, McGraw-Hill, New York.

P

PARTITION COEFFICIENTS: Octanol–Water (K_{ow})

Definition

The octanol–water partition coefficient is the ratio of the concentration of an organic compound in an organic phase (e.g., octanol) to its concentration in water.

Formula

$$K_{ow} = \frac{C_o}{C_w}$$

where

C_o = concentration of compound in octanol (mol/L)

C_w = concentration of compound in water (mol/L)

In environmental toxicology, this partition coefficient is useful for predicting the bioconcentration potential of hydrophobic compounds in the biota and their sorption by organic adsorbents such as organic matter of soils and sediments.

Numerical Values

See Table P1.

References

Donnely, K. C., C. S. Anderson, G. C. Barbee, and D. J. Manek. 1994. Soil toxicology, pp. 321–352. In: *Basic Environmental Toxicology*, L. G. Cockerham and B. S. Shane, Eds., CRC Press, Boca Raton, FL.

Schnoor, J. L. et al. 1987. *Processes, Coefficients, and Models for Simulating Toxic Organics and Heavy Metals in Surface Waters*, EPA-600/3-87/015. U.S. Environmental Protection Agency, Washington, DC.

Schwarzenbach, R. P., P. M. Gschwend, and D. M. Imboden. 1993. *Environmental Organic Chemistry*, Wiley-Interscience, New York.

TABLE P1: S_w, K_{ow}, and K_{oc} of Some Organic Chemicals[a]

Class	Constituent	S_w^a (mg/L)	K_{oc} (mL/g)	Log K_{ow}
Chlorinated organics	Trichloroethylene	1,100	126	2.38
	PCBs		5.3×10^5	6.04
	Chloroform	8,200	31	1.97
	Tetrachloroethylene	200	364	2.60
	1,1,1-Trichloroethane	720	152	2.50
	Methylene chloride	20,000	9	1.30
	trans-1,2-Dichloroethylene	600	59	0.48
	1,2-Dichloroethane	8,690	14	1.48
	Vinyl chloride	1.1	57	1.38
	Chlorobenzene	488	330	2.84
	1,2-Dichlorobenzene	100		3.56
	1,1-Dichloroethane	5,500	30	1.79
	Carbon tetrachloride	785	110	2.64
	Pentachlorophenol	14		5.04
	Chloromethane	6,450		0.95
Aromatic organics	Toluene	515	300	2.73
	Benzene	1,780	83	2.12
	Naphtalene	31		3.29
	Phenol	93,000	14	1.46
	Ethybenzene	152	1100	3.15
	Xylene			
	ortho	175	363	3.04
	meta	130	588	3.20
	para	198	552	3.16
	Nitrobenzene	1,900		1.87
	2-Nitrophenol	2,100		1.75
Miscellaneous compounds	DDT	0.0055		6.91
	2,4-D	900		1.78
	Atrazine	33		2.69
	Lindane	7.52		3.72
	Dieldrin	0.2		3.54

[a] S_w, water solubility; K_{oc}, organic carbon normalized soil sorption coefficient; K_{ow}, octano–water partition coefficient.

Source: Adapted from Donnely et al. (1994); Schnoor et al. 1987.

PARTITION COEFFICIENT: Soil Organic Matter/Soil Water (K_{oc})

Introduction

Soil organic matter (e.g., humic substances) is the main adsorbent of hydrophobic organic chemicals in soils. Thus we use the partition coefficient K_{oc} to describe the distribution of organic chemicals between soil pore water and soil organic matter.

Formulas

$$K_{oc} = \frac{C_{oc}}{C_{pw}}$$

where

C_{oc} = chemical concentration in the organic matter (mg chemical/g soil organic carbon)

C_{pw} = chemical concentration in the pore water (mg chemical/ml water)

K_{oc} can be obtained by dividing the solid/water distribution coefficient K_d by the mass fraction of organic carbon in the solid f_{oc}:

$$K_{oc} = \frac{K_d}{f_{oc}}$$

For nonpolar chemicals, there is a significant negative correlation between K_{oc} and water solubility S_w (Hasset et al., 1983):

$$\log K_{oc} = 3.95 - 0.62 \log S_w$$

Another equation reported for hydrophobic compounds is (Karickhoff et al., 1979)

$$\log K_{oc} = 0.44 - 0.54 \log S_w$$

Ryan et al. (1988) have summarized the relationship between K_{oc} and K_{ow} as shown by the following equations from Means et al. (1982), Schwarzenbach and

Westall (1981), Rao et al. (1982), Karickhoff (1981), and Brown and Flagg (1981), respectively:

$$\log K_{oc} = \log K_{ow} - 0.317$$
$$\log K_{oc} = 0.721 \log K_{ow} + 0.49$$
$$\log K_{oc} = 1.029 \log K_{ow} - 0.18$$
$$\log K_{oc} = 0.989 \log K_{ow} - 0.346$$
$$\log K_{oc} = 0.937 \log K_{ow} - 0.006$$

Numerical Values

See Table P1.

References

Brown, D. S., and E. W. Flagg. 1981. Empirical prediction of organic pollutant sorption in natural sediments. J. Environ. Qual. 10: 382–386.

Donnely, K. C., C. S. Anderson, G. C. Barbee, and D. J. Manek. 1994. Soil toxicology, pp. 321–352. In: *Basic Environmental Toxicology*, L. G. Cockerham and B. S. Shane, Eds. CRC Press, Boca Raton, FL.

Hasset, J. J., W. L. Banwart, and R. A. Griffin. 1983. Correlation of compound properties with sorption characteristics of nonpolar compounds by soils and sediments: concepts and limitations, pp. 161–178. In: *Environment and Solid Wastes: Characterization, Treatment and Disposal*, E. C. Francis and S. Auerbach, Eds. Butterworth, Newton, MA.

Karickhoff, S. W. 1981. Semiempirical estimation of sorption of hydrophobic pollutants on natural sediments and soils. Chemosphere 10: 833–846.

Karickhoff, S. W., D. S. Brown, and T. A. Scott. 1979. Sorption of hydrophobic pollutants on natural sediments. Water Res. 13: 241–248.

Means, J. C., S. G. Wood, J. J. Hassett, and W. L. Banwart. 1982. Sorption of amino and carboxy-substituted polynuclear aromatic hydrocarbons by sediments and soils. Environ. Sci. Technol. 16: 93–98.

Rao, P. S. C., J. M. Davidson, V. E. Berkhiser, and L. T. Ou. 1982. *Retention and Transformation of Selected Pesticides and Phosphorus in Soil Water Systems: A Critical Review*, EPA-660/3-83-060, U.S. Environmental Protection Agency, Washington, DC.

Ryan, J. A., R. M. Bell, J. M. Davidson, and G. A. O'Connor. 1988. Plant uptake of nonionic organic chemicals from soils. Chemosphere 17: 2299–2323.

Schwarzenbach, R. P., and J. Westall. 1981. Transport of nonpolar organic compounds from surface water to groundwater. Environ. Sci. Technol. 15: 1360–1367.

PARTITION COEFFICIENT: Soil Water/Air

Introduction

The soil water/air distribution coefficient describes the distribution of a chemical between soil pore water and soil atmosphere.

Formula

In soils, the distribution of an organic chemical between the soil pore water and air is given by the following formula (Donnely et al., 1994; Ryan et al., 1988):

$$H_c = \frac{16.04 V_p M}{T S_w}$$

where

H_c = Henry's constant ($\text{atm} \cdot \text{m}^3/\text{mol}$)
V_p = vapor pressure of pure chemical (atm)
M = molecular weight (g/mol)
T = absolute temperature (K)
S_w = water solubility of chemical (g/m^3)

The reciprocal of H_c is known as the *air/water partition coefficient* K_w. K_w is expressed as (μg chemical/mL)/(μg chemical/cm^3 air). Chemicals with low K_w volatilize from soils.

References

Donnely, K. C., C. S. Anderson, G. C. Barbee, and D. J. Manek. 1994. Soil toxicology, pp. 321–352. In: *Basic Environmental Toxicology*, L. G. Cockerham and B. S. Shane, Eds. CRC Precc, Boca Raton, FL.

Ryan, J. A., R. M. Bell, J. M. Davidson, and G. A. O'Connor. 1988. Plant uptake of nonionic organic chemicals from soils. Chemosphere 17: 2299–2323.

PARTITION COEFFICIENT: Solid/Water (K_d)

Introduction

The partition coefficient K_d describes the distribution of a chemical between a soil or sediment and the soil or sediment pore water at equilibrium. It is determined by mixing a soil or sediment with a chemical for a given time period (batch study) or by measuring the transport of the chemical through saturated soil columns.

Formula

$$K_d = \frac{C_s}{C_{pw}}$$

where

K_d = partition or distribution coefficient (mg chemical/kg solid)/(mg chemical/liter water), or as L water/kg solid). (the slope of the adsorption isotherm)

C_s = mg chemical/kg soil or sediment

C_{pw} = chemical concentration in the pore water (mg chemical/L)

The equation above assumes complete reversibility at equilibrium between the solid and the liquid phases (Ryan et al., 1988). K_d can be useful for obtaining the retardation factor for groundwater contaminants (*see* RETARDATION FACTOR: Groundwater Contaminants).

References

Donnely, K. C., C. S. Anderson, G. C. Barbee, and D. J. Manek. 1994. Soil Toxicology, pp. 321–352. In: *Basic Environmental Toxicology*, L. G. Cockerham and B. S. Shane, Eds. CRC Press, Boca Raton, FL.

Ryan, J. A., R. M. Bell, J. M. Davidson, and G. A. O'Connor. 1988. Plant uptake of nonionic organic chemicals from soils. Chemosphere 17: 2299–2323.

pH

Definition

In 1909, Sorensen proposed the expression of hydrogen ion concentration as pH or the negative logarithm of the hydrogen ion concentration.

Formula

$$\mathrm{pH} = \log_{10}\frac{1}{[\mathrm{H}^+]} = -\log_{10}[\mathrm{H}^+]$$

Water is acidic at $\mathrm{pH} < 7$ and basic at $\mathrm{pH} > 7$.

Reference

Sawyer, C. N., P. L. McCarty, and G. F. Parkin. 1994. *Chemistry for Environmental Engineering*, McGraw-Hill, New York.

PHELPS EQUATION

See TEMPERATURE COEFFICIENT

Introduction

Phelps (1944) proposed an expression for quantifying temperature effects on reaction rates. This expression can be derived from the Arrhenius equation at a specific temperature (*see* ARRHENIUS EQUATION).

Formula

$$K_1 = K_2 \theta^{T_1 - T_2}$$

where

K_1 = reaction rate coefficient at temperature T_1

K_2 = reaction rate coefficient at temperature T_2 (T_2 normally chosen as 20°C)

θ = thermal coefficient (obtained by plotting ln K versus temperature)

References

Grady, C. P. L., and H. C. Lim. 1990. *Biological Wastewater Treatment: Theory and Applications*, Marcel Dekker, New York.

Phelps, E. B. 1994. *Stream Sanitation*, Wiley, New York.

PHENOL COEFFICIENT

Definition/Introduction

The phenol coefficient of a given disinfectant is the ratio of the dilution of the disinfectant that kills specific microorganisms in 10 min, to the dilution of phenol that results in a similar effect. Specific known bacterial strains used for determining the phenol coefficient are *Pseudomonas aeruginosa*, *Staphylococcus aureus*, and *Salmonella typhi*.

TABLE P2: Phenol Coefficients of Some Chlorinated Compounds (Using *S. typhi*)

Compound	Phenol Coefficient
Phenol	1
meta-Cresol	2
Hexachlorophene	15
para-Chloro-*meta*-cresol	11
para-*n*-amyl-*ortho*-chlorophenol	80

Formula

For a given disinfectant:

$$\text{Phenol coefficient} = \frac{\text{maximum dilution of disinfectant}}{\text{maximum dilution of phenol that results in the same effect}}$$

For example the dilution of *para*-chlorophenol that kills *Salmonella typhi* in 10 minutes = 1:360; the dilution of phenol that kills *S. typhi* in 10 minutes = 1:90. Then

$$\text{phenol coefficient} = \frac{360}{90} = 4$$

Numerical Values

See Table P2.

Reference

Nester, E. W., N. N. Pearsall, J. B. Roberts, and C. E. Roberts. 1982. *The Microbial Perspective*, Saunders College Publishing, Philadelphia.

PHOSPHORUS: Critical Loading in Lakes

Introduction

When present in relatively excessive amounts in water, nitrogen and phosphorus cause lake eutrophication. The trophic status is influenced by the amounts of nutrients entering lakes. As regards phosphorus, a given lake may pass from oligotrophic to eutrophic conditions when this nutrient exceeds a certain critical loading value (Cole, 1979; Vollenweider, 1976).

Formula

The annual critical loading value for phosphorus in lakes is given by

$$L_c = (10 \text{ to } 20) \times q_s(1 + \sqrt{z/q_s})$$

where

L_c = annual critical loading value (mg P/m^2)
q_s = hydraulic load (m/year)
z = mean depth (m)

References

Cole, G. A. 1979. *Textbook of Limnology*, C.V. Mosby, St. Louis, MO.

Vollenweider, R. A. 1976. Advances in defining critical loading levels for phosphorus in lake eutrophication. Mem. Ist. Ital. Idrobiol. 17: 191–206.

PHOTON: Energy

See ENERGY OF A PHOTON.

PHOTOSYNTHESIS: *P/R* Ratio

Definition

The P/R ratio is the ratio of total or gross primary production (*P*) over total respiration (*R*). This ratio is used to classify communities according to their autotrophic or heterotrophic characteristics.

Formula

$$P/R = \frac{\text{total primary production (g/m}^2 \cdot \text{day)}}{\text{total community respiration (g/m}^2 \cdot \text{day)}}$$

Numerical Values

P/R ratio is close to 1 in oligotrophic lakes and is substantially greater than 1 in eutrophic lakes. In aquatic environments receiving wastewater (i.e., input of organic substances), P/R can decrease significantly below 1. P/R was reported as low as 0.008 for heterotrophic communities (White River, Indiana, near a sewage outfall) to 7.0 for autotrophic communities (Silver Springs, Florida) (Odum, 1956).

References

Cole, G. A. 1988. *Textbook of Limnology*, 3rd ed., Waveland Press, Prospect Heights, IL.

Odum, H. T. 1956. Primary production in flowing waters. Limnol. Oceanogr. 1: 102–117.

PHOTOSYNTHETIC EFFICIENCY (PE)

Definition

Photosynthetic efficiency (PE) is the efficiency of conversion of the energy in incident sunlight to gross primary production by primary producers.

Formula

Photosynthetic efficiency (PE) is represented as follows (Krebs, 1972):

$$\text{PE} = \frac{\text{energy fixed by gross primary production}}{\text{energy in incident sunlight}}$$

$$\text{PE (\%)} = \frac{P \times 2 \times 5500}{\text{solar energy} \times 0.5} \times 100$$

where

P = primary productivity ($\text{g C/m}^2 \cdot \text{day}$)

2 = conversion factor for carbon to dry algal mass

5500 = approximate caloric equivalent/g dry algal biomass

0.5 = conversion of total solar radiation to the portion of spectrum that is photosynthetically active (PAR)

and where solar energy is expressed in $\text{cal/m}^2 \cdot \text{day}$.

Numerical Values

In general, the efficiency of gross primary production is less than 2% (Krebs, 1972). For example, the efficiency of the aquatic community of Lake Mendota, Wisconsin, was reported as 0.42% (Kozlovsky, 1968). A 4% PE was observed in eutrophic lakes (Wetzel, 1983).

References

Kozlovsky, D. G. 1968. A critical review of the trophic level concept. I. Ecological efficiencies. Ecology 49: 48–60.

Krebs, C. J. 1972. *Ecology*, Harper & Row, New York.

Lind, O. W. 1979. *Handbook of Common Methods in Limnology*, 2nd ed., C.V. Mosby, St. Louis, MO.

Wetzel, R. G. 1983. *Limnology*, 2nd ed., Saunders College Publishing, Philadelphia.

Wetzel, R. G., and G. E. Likens. 1991. *Limnological Analyses*, 2nd ed., Springer-Verlag, New York.

PHOTOSYNTHETIC QUOTIENT (PQ)

Introduction

It is now customary to express primary production in aquatic environments in terms of carbon fixed rather than oxygen produced as a result of photosynthesis. The photosynthetic quotient PQ allows the conversion of O_2 produced to CO_2 fixed.

Formula

$$PQ = \frac{O_2 \text{ produced}}{CO_2 \text{ fixed}}$$

Numerical Values

A photosynthetic quotient of 1.2 is generally used. For conversion from O_2 to C, one uses a conversion factor of 0.375 (0.375 = atomic weight of C/molecular weight of oxygen = 12/32) (Cole, 1979). For example, if the annual production of oxygen is $0.5\,\text{kg/m}^2$ and if PQ = 1.2, the equivalent carbon is

$$0.5 \text{ kg } O_2 \times 0.375 \times 1/1.2 = 0.156 \text{ kg C}$$

References

Cole, G. A. 1979. *Textbook of Limnology*, C.V. Mosby, St. Louis, MO.

Raymont, J. E. G. 1980. *Plankton and Productivity in the Oceans*, vol. 1, *Phytoplankton*, 2nd ed., Pergamon Press, Oxford.

PHOTOSYNTHETIC RATE: Relationship with Chlorophyll

Introduction

It was suggested that the photosynthesis rate in water is related to chlorophyll *a* concentration (U.S. EPA, 1973).

Formula

$$P = 0.25[\mathrm{CHl}_a]$$

where

P = optimum rate of photosynthesis (mg/L · day)

$[\mathrm{Chl}_a]$ = chlorophyll *a* concentration (mg/L)

References

Liu, C. C. K. 1986. Surface water quality analysis. In: *Handbook of Environmental Engineering*, vol. 4, *Water Resources and Natural Control Processes*, L. K. Wang and N. C. Pereira, Eds., Humana Press, Clifton, NJ.

U.S. EPA. 1973. *Guidelines for Developing or Revising Water Quality Standards Under the Water Pollution Control Act Amendments of 1972*. Water Planning Division, Washington, DC.

PLANCK EQUATION

Introduction

The Planck equation states that the energy of a photon is directly related to light frequency and inversely related to wavelength.

Formula

$$\varepsilon = h\nu$$

The light frequency ν is related to wavelength by the following:

$$\nu = \frac{c}{\lambda}$$

Thus

$$\varepsilon = h\nu = \frac{hc}{\lambda}$$

where

ε = energy of photon (J)

h = Planck's constant = 6.625×10^{-34} J · s

ν = light frequency (s^{-1})

c = speed of light = 3×10^8 m/s

λ = wavelength (m)

References

Cole, G. A. 1979. *Textbook of Limnology*, C.V. Mosby, St. Louis, MO.

Tinoco, I., Jr., K. Sauer, and J. C. Wang. 1995. *Physical Chemistry: Principles and Applications in Biological Sciences*, Prentice Hall, Upper Saddle River, NJ.

PONDS

See COLIFORM DIE-OFF KINETICS; GAS PRODUCTION IN FACULTATIVE PONDS; METHANE PRODUCTION; OXYGEN PRODUCTION.

POPULATION EQUIVALENTS

Introduction

Coliform densities measured in a stream can be converted to population equivalents (PE), based on the finding that coliform contribution is around 200×10^9/capita · day (yearly average).

Formula

$$PE = \frac{\text{coliform mean density}/100 \text{ ml} \times Q}{C \times 100}$$

where

Q = stream flow (m^3/s)

$$C = \frac{200 \times 10^9 (\text{coliforms/capita} \cdot \text{day})}{86.4 \times 10^9} = 2.315$$

1 m^3/s stream flow = 86.4×10^9 mL/day.

Reference

Velz, C. 1984. *Applied Stream Sanitation*, Wiley, New York.

POPULATION GROWTH

See GROWTH POPULATION.

POROSITY, SOIL

Definition

The porosity of a soil is the percentage of the total soil volume that is occupied by pores. The pores are filled with water in saturated soils or with air and water in unsaturated soils.

Formula

$$\varepsilon = \frac{V_t - V_s}{V_t}$$

or

$$\varepsilon = \frac{V_l + V_g}{V_t}$$

where

- ε = porosity (dimensionless)
- V_t = total volume in undisturbed sample
- V_s = volume of the solid phase (obtained by oven drying the soil and deriving the volume V_s from dry weight and soil density)
- V_l = volume of the liquid phase
- V_g = volume of the gas phase

Porosity is also expressed in percentage as

$$n = \frac{V_l + V_g}{V_t} \times 100$$

The percentage of pore space in a given volume of soil is also given by the following:

$$\% \text{ pore space} = 100\% - \% \text{ solid space}$$

or

$$\% \text{ pore space} = 100\% - \left(\frac{\text{bulk density}}{\text{particle density}} \times 100\right)$$

For soil bulk density, *see* BULK DENSITY OF SOILS.

In soil mechanics, the pore volume of a soil is expressed as *void ratio e*, which is given by the following formula:

$$e = \frac{V_v}{V_s}$$

where

V_v = volume of the voids ($= V_l + V_g$)

V_s = volume of the solids

Porosity and void ratio are related by the following formulas:

$$\varepsilon = \frac{e}{e + 1} \quad \text{or} \quad e = \frac{\varepsilon}{1 - \varepsilon}$$

Numerical Values

See Table P3.

References

Bouwer, H. 1978. *Groundwater Hydrology*, McGraw-Hill, New York.

Davis, S. N. 1969. Porosity and permeability of natural materials, pp. 54–89. In: *Flow Through Porous Media*, R. J. M. de Wiest, Ed., Academic Press, San Diego, CA.

Donahue, R. L., R. W. Miller, and J. C. Shikluma. 1977. *Soils: An Introduction to Soils and Plant Growth*. Prentice Hall, Upper Saddle River, NJ.

Freeze, R. A., and J. A. Cherry. 1979. *Groundwater*, Prentice Hall, Upper Saddle River, NJ.

Marshall, T. J., and J. W. Holmes. 1988. *Soil Physics*, 2nd ed., Cambridge University Press, Cambridge.

TABLE P3: Porosity of Soils and Soil Constituents

Material	Porosity (%)
Gravel	25–40
Sand	25–50
Silt	35–50
Clay	40–70
Surface soil of wet clay	58
Decomposed peat	~65
Surface soil of a loamy sand	43
Subsoil of sandy texture	39
Sandy loam soil compacted by heavy traffic	28

PREDATION IN AN OPEN ECOSYSTEM

Introduction

Both the predator and prey grow together in an open ecosystem and follow Monod's kinetics.

Formulas

$$\text{Predator}: \quad \frac{dP}{dt} = \frac{\lambda_{\max}\mathrm{NP}}{L_s + N} - \mathrm{DP}$$

$$\text{Prey}: \quad \frac{dN}{dt} = \frac{\mu_{\max}SN}{K_s + S} - \frac{\lambda_{\max}NP}{W(L_s + N)}$$

where

N = prey concentration
$\mu_{\max}$ = maximum specific growth rate of prey (h^{-1})
S = substrate concentration (mg/L)
K_s = half saturation constant for prey (mg/L)
$\lambda_{\max}$ = maximum specific growth rate of predator (h^{-1})
P = predator concentration
W = yield of predator/unit of prey consumed
L_s = half saturation constant for predator (mg/L)
D = dilution rate (H^{-1})

According to the Lotka–Volterra model, the relationship between prey and predator is described by the following equations (Smith, 1996):

$$\text{Prey}: \quad \frac{dN}{dt} = aN - bNP$$

$$\text{Predator}: \quad \frac{dP}{dt} = cNP - dP$$

where

N = prey density
P = predator density
a = rate of change of prey in the absence of predator
b = rate of change of prey in the presence of predator
c = rate of change of predator in the presence of prey
d = rate of change of predator in the absence of prey

References

Bazin, M., and A. Menell. 1990. Mathematical methods in microbial ecology. Methods Microbiol. 22: 125–179 (R. Grigorova and J. R. Norris, Eds.), Academic Press, London.

Smith, R. L. 1996. *Ecology and Field Biology*, 5th ed., Harper Collins, New York.

PRESSURE, ATMOSPHERIC

Formulas

The atmospheric pressure P_h at any altitude h is given by

$$\log P_h = \log P_0 - \frac{273h}{18{,}421 q_K}$$

where

P_h = pressure at altitude h
P_0 = pressure at sea level (760 mm Hg)
h = altitude (m)
q_K = mean absolute temperature of the air column between sea level and the level h (K)

Meteorological experience indicates that an increase of 200 m in elevation generally is accompanied by a decrease in mean temperature of 1°C. Hence the pressure can be approximated further by

$$\log P_h = \log P_0 - \frac{273h}{18{,}421(q_0 + 273 + 0.0025h)}$$

where q_0 is the absolute temperature at the water surface (K).

Reference

Hutchinson, G. E. 1975. *A Treatise on Limnology*, vol. I, *Geography, Physics, and Chemistry*, Part 2 *Chemistry of Lakes*, Wiley, New York.

PRIMARY PRODUCTIVITY

Introduction

Primary productivity is the rate of formation by primary producers of organic matter over a defined period of time. Planktonic algal photosynthesis can be calculated, using one of two methods: oxygen and ^{14}C methods.

Oxygen Method

Oxygen changes due to plankton photosynthesis, and respiration are measured after 24 h or less in dark (D) and light (L) bottles containing the sample under investigation.

Formulas

$$\text{Respiratory activity (mg O}_2\text{ consumed/L}\cdot\text{h)} = \frac{I - D}{t}$$

$$\text{Net photosynthetic activity (mg O}_2\text{/L}\cdot\text{h)} = \frac{L - I}{t}$$

$$\text{Gross photosynthetic activity (mg O}_2\text{/L}\cdot\text{h)} = \frac{L - D}{t}$$

where

I = initial oxygen concentration (mg/L)

L = oxygen concentration (mg/L) in light bottle

D = oxygen concentration (mg/L) in dark bottle

t = incubation period (h)

The gross primary productivity in mg $C/m^3 \cdot$ day is (Coler and Rockwood, 1989)

$$\text{gross primary productivity} = \frac{L - D}{t} \times \frac{12}{32} \times 1000 \times 12$$

where

12/32 = atomic weight of carbon/molecular weight of oxygen
1000 = conversion factor for liters to m^3
12 = hypothetical number of hours of light per day

^{14}C Method

Use light and dark bottles inoculated with $Na_2\,{}^{14}CO_3$. Incubate at desired depth. Retrieve bottles, preserve, and transport to the laboratory. In the laboratory, filter sample or an aliquot of the sample. Measure radioactivity of filter with a scintillation counter.

Formulas

$$P = P_l - P_d$$

where

P = photosynthesis (mg C/m^3)
P_l = C uptake/m^3 in light bottle
P_d = carbon uptake/m^3 in dark bottle

$$P_l = \frac{r}{R \times C \times f}$$

where

r = uptake of radioative C (cpm)
$= \text{cpm for filtered volume} \times \frac{\text{volume in bottle}}{\text{volume filtered}}$
R = ^{14}C added to bottle (cpm) = 2.2×10^6 microcuries added × counter efficiency
C = total inorganic ^{12}C present in sample in mg/m^3
= total alkalinity × 1000 × conversion factor of total alkalinity to mg of carbon per liter (see Table P4)
f = correction factor (isotope discrimination factor) for slower uptake (6% slower) of ^{14}C as compared to $^{12}C = 1.06$

TABLE P4: Factors for Conversion of Total Alkalinity to Milligrams of Carbon per liter

	Temperature (°C)					
pH	0	5	10	15	20	25
6.0	1.15	1.03	0.93	0.87	0.82	0.78
6.1	0.96	0.87	0.77	0.73	0.70	0.67
6.2	0.82	0.74	0.68	0.64	0.60	0.58
6.3	0.69	0.64	0.59	0.56	0.53	0.51
6.4	0.60	0.56	0.52	0.49	0.47	0.45
6.5	0.53	0.49	0.46	0.44	0.42	0.41
6.6	0.47	0.44	0.41	0.40	0.38	0.37
6.7	0.42	0.40	0.38	0.37	0.35	0.35
6.8	0.38	0.37	0.35	0.34	0.33	0.32
6.9	0.35	0.34	0.33	0.32	0.31	0.31
7.0	0.33	0.32	0.31	0.30	0.30	0.29
7.1	0.31	0.30	0.29	0.29	0.29	0.28
7.2	0.30	0.29	0.28	0.28	0.28	0.27
7.3	0.29	0.28	0.27	0.27	0.27	0.27
7.4	0.28	0.27	0.27	0.26	0.26	0.26
7.5	0.27	0.26	0.26	0.26	0.26	0.26
7.6	0.27	0.26	0.26	0.25	0.25	0.25
7.7	0.26	0.26	0.25	0.25	0.25	0.25
7.8	0.25	0.25	0.25	0.25	0.25	0.25
7.9	0.25	0.25	0.25	0.25	0.25	0.25
8.0	0.25	0.25	0.25	0.25	0.24	0.24
8.1	0.25	0.25	0.24	0.24	0.24	0.24
8.2	0.24	0.24	0.24	0.24	0.24	0.24
8.3	0.24	0.24	0.24	0.24	0.24	0.24
8.4	0.24	0.24	0.24	0.24	0.24	0.24
8.5	0.24	0.24	0.24	0.24	0.24	0.24
8.6	0.24	0.24	0.24	0.24	0.24	0.24
8.7	0.24	0.24	0.24	0.24	0.24	0.24
8.8	0.24	0.24	0.24	0.24	0.23	0.23
8.9	0.24	0.24	0.23	0.23	0.23	0.23

Source: Saunders et al. (1962).

Numerical Values

- *Primary productivity rates of phytoplankton in fresh waters:* see Table P5.
- *Gross Primary Productivity of some aquatic and terrestrial environments*: see Table P6.
- *Net primary productivity in limnic ecosystems:* see Table P7.

TABLE P5: Primary Productivity Rates of Phytoplankton in Fresh Waters

Trophic Status	Range of Daily Productivity for Entire Year (mg C/m^2 · day)
Oligotrophic	1.6–249
Mesotrophic	210–729
Eutrophic	1012–1750

Source: Adapted from Wetzel (1983).

Table P6: Gross Primary Productivity of Estuaries as Compared to Other Aquatic and Terrestrial Areas

Site	Gross Primary Production[a] (kcal/m^2 · y)
Estuaries and reefs	20,000
Tropical forest	20,000
Fertilized farmland	12,000
Eutrophic lakes	10,000
Unfertilized farmland	8,000
Coastal upwellings	6,000
Grassland	2,500
Oligotrophic lakes	1,000
Open oceans	1,000
Desert and tundra	200

Source: Adapted from Goldman and Horne (1983).

[a] Net carbon production was converted to gross energy values, where necessary, by multiplying by 20.

References

Coler, R. A., and J. P. Rockwood. 1989. *Water Pollution Biology*, Technomic, Lancaster, PA.

Goldman, C. R., and A. J. Horne. 1983. *Limnology*, McGraw-Hill, New York.

Lind, O. W. 1979. *Handbook of Common Methods in Limnology*, 2nd ed., C.V. Mosby, St. Louis, MO.

Saunders, G. W., F. B. Trama, and R. W. Bachmann. 1962. Evaluation of a modified C-14 technique for shipboard estimation of photosynthesis in large lakes. Great Lakes Res. Div. Publ. 8, University of Michigan, Ann Arbor, MI.

Schwoerbel, J. 1987. *Handbook of Limnology*, Ellis Horwood, Chichester, West Sussex, England.

Wetzel, R. G. 1983. *Limnology*, Saunders College Publishing, Philadelphia.

TABLE P7: Net Primary Productivity in Limnic Ecosystems

Ecosystem	$mg\,C/m^2 \cdot day$
Tropical lakes	100–7600
Temperate lakes	3–3600
Arctic lakes	1–170
Antarctic lakes	1–35
Alpine lakes	1–450
Temperate streams	<1–3000
Oligotrophic lakes	50–300
Mesotrophic lakes	250–1000
Eutrophic lakes	600–8000
Dystrophic lakes	<50–500

Source: Adapted from Schwoerbel (1987).

PROBIT UNITS

Introduction

The response of a test organism to a toxic chemical is represented by a sigmoid curve (Figure M1) (see MEDIAN LETHAL CONCENTRATION). It is difficult to determine the LC_{50} from the sigmoid curve. In a normally distributed population the response of the test organism can be linearized by converting the organism response (e.g., death) to units of deviation from the mean named *normal equivalent deviations* (NED). A 50% response corresponds to an NED of 0, while a 15.1% response corresponds to an NED of -1.

Formula

Bliss (1957) suggested the probits units to avoid the negative numbers of NED. Probits units are related to NED by

$$\text{probit unit} = \text{NED} + 5$$

The relationship between percent response, NED, and probit units is given in Table P8. Determination of the LC_{50} using probit units is shown in Figure P1.

TABLE P8: Relationship Between Percent Response, Normal Equivalents Deviations (NED) and Probit Units

% Response	NED	Probit Units
0.1	−3	2
2.3	−2	3
15.9	−1	4
50	0	5
84.1	+1	6
97.7	+2	7
99.9	+3	8

Source: Shane (1994).

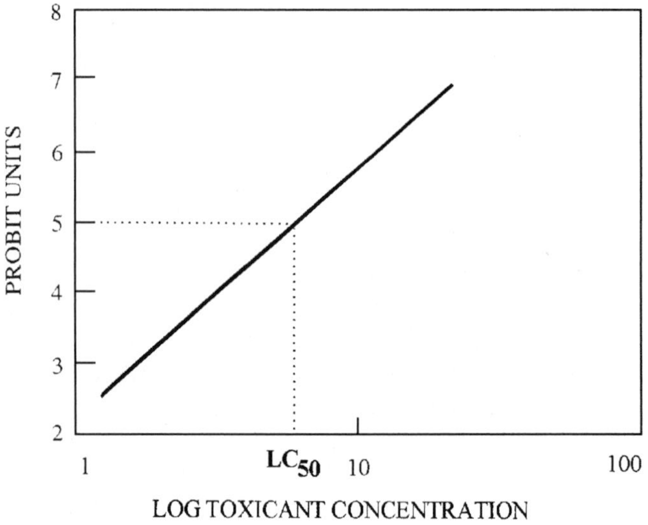

FIGURE P1: Determination of LC_{50} in Toxicity Testing via the Probit Units Method.

References

Bliss, C. L. 1957. Some principles of bioassay, Am. Sci., 45: 449–466.

Shane, B. S. 1994. Principles of ecotoxicology, pp. 11–47. In: *Basic Environmental Toxicology*, L. G. Cockersham and B. S. Shane. Eds. CRC Press, Boca Raton, FL.

PRODUCTION

See BACTERIAL PRODUCTION.

PRODUCTIVITY

See PRIMARY PRODUCTIVITY

PRODUCTIVITY/BIOMASS (*P*/*B*) RATIO

Definition

The productivity/biomass (P/B) ratio is the ratio of annual production to mean annual biomass. P/B ratio is used to estimate the turnover rate of organisms, and decreases with increasing trophic level.

Numerical Values

See Table P9.

References

Brylinsky, M. 1980. Estimating the productivity of lakes and reservoirs, pp. 441–453. In: *The Functioning of Freshwater Ecosystems*, E. D. Le Cren and R. H. Lowe-McConnell, Eds., Cambridge University Press, Cambridge.

Saunders, G. W., K. W. Cummins, D. Z. Gak, E. Pieczynska, V. Straskrabova, and R. G. Wetzel. 1980. Organic matter and decomposers, pp. 341–392. In: *The Functioning of Freshwater Ecosystems*, E. D. Le Cren and R. H. Lowe-McConnell, Eds., Cambridge University Press, Cambridge.

Wetzel, R. G. 1983. *Limnology*, 2nd ed., Saunders College Publishing, Philadelphia.

TABLE P9: Productivity/Biomass (P/B) Ratios of Freshwater Organisms

Organism	Mean P/B ratio
Bacteria	141.0
Algae	113.0
Herbivorous zooplankton	15.9
Carnivorous zooplankton	11.6
Herbivorous benthic invertebrates	3.7
Carnivorous benthic invertebrates	4.8

PROTON MOTIVE FORCE

Definition/Introduction

The movement of protons across biological membranes is coupled with ATP synthesis. The proton motive force that propels the protons is the sum of proton concentration gradient and membrane electric potential. In mitochondria, the membrane electric potential (approximately 200 mV) is a major component of the proton motive force.

Formula

$$\Delta p = \Delta\Psi - 2.3\frac{RT}{F}(\Delta\text{pH})$$

where

Δp = proton motive force (mV)
$\Delta\Psi$ = transmembrane electric potential (mV)
R = gas constant
T = temperature (K)
F = Faraday constant
ΔpH = difference between interior and exterior of the cell

At 25°C:

$$2.3\frac{RT}{F} = 59 \text{ mV}$$

The membrane potential is estimated by measuring the distribution of an inorganic (e.g., K^+) or organic (e.g., fluorescent dyes) ion to which the cell is permeable.

References

Darnell, J., H. Lodish, and D. Baltimore. 1990. *Molecular Cell Biology*, Scientific American, New York.

Maloney, P. C., E. R. Kashket, and T. H. Wilson. 1975. Methods for studying transport in bacteria. Methods Membrane Biol. 5: 1–49.

Marquis, R. E. 1981. Permeability and transport, pp. 393–404. In: *Manual of Methods for General Bacteriology*, P. Gerhardt, R. G. E. Murray, R. N. Costilow, E. W. Nester, W. A. Wood, N. R. Krieg, and C. B. Phillips, Eds. American Society for Microbiology, Washington, DC.

PROTOZOA

See CARBON CONVERSION EFFICIENCY: Protozoa; CLEARANCE RATE: Zooplankton and Protozoa.

Q

Q_{10}

Definition

The van't Hoff rule states that the reaction rate of an enzyme doubles as the temperature increases by 10°C. The Q_{10} value expresses change in the activity of an enzyme as a result of a 10°C increase in temperature within the tolerance range of the enzyme.

Formula

$$Q_{10} = \frac{k_{T+10}}{k_T}$$

where

k_T = rate constant at temperature T

K_{T+10} = rate constant at temperature $T + 10$°C

Numerical Values

Q_{10} is generally approximately 2. Q_{10} values for some enzymes are shown in Table Q1.

TABLE Q1: Q_{10} for Selected Enzymes

Enzyme	Temperature (°C)	Q_{10}
Catalase	10–20	2.2
Urease	20–30	1.8
Maltase	10–20	1.9
	20–30	1.4
Succinic oxidase	30–40	2.0
Amylase	10–20	1.3
	20–30	1.4
Pepsin	10–20	2.0
	20–30	1.8
Xanthine oxidase	27–37	4.0

Source: Data from Atlas and Bartha (1981), Christensen and Palmer (1974), Sawyer et al. (1984).

References

Atlas, R. M., and R. Bartha. 1981. *Microbial Ecology: Fundamentals and Applications*, Adison-Wesley, Reading, MA.

Christensen, H. N., and G. A. Palmer. 1974. *Enzyme Kinetics*, 2nd ed., W.B. Saunders, Philadelphia.

Sawyer, C. N., P. L. McCarty, and G. F. Parkin. 1994. *Chemistry for Environmental Engineering*, McGraw-Hill, New York.

Williams, V. R., W. L. Mattice, and H. B. Williams. 1978. *Basic Physical Chemistry for the Life Sciences*, W.H. Freeman, San Francisco.

R

RADIOACTIVE DECAY

See also REACTION KINETICS.

Introduction

Radioactive substances undergo decay, following first-order kinetics.

Formula

Radioactivity remaining after time t is given by the following equation:

$$N_t = N_0 e^{-\lambda t}$$

where

$$\lambda = \frac{0.693}{t_{1/2}}$$

and where

N_t = number of radioactive atoms at time t
N_0 = number of radioactive atoms at time 0
$t_{1/2}$ = half-life of radioactive substance

Reference

Efiok, B. J. S. 1993. *Basic Calculations for Chemical and Biological Analyses*, AOAC International, Arlington, VA.

REACTION KINETICS: First-Order Reaction

See also CHICK'S LAW.

Introduction

Many reactions of importance to environmental engineers and scientists follow first-order kinetics. Examples are pathogen decay following disinfection, organic matter degradation in water and wastewater (*see* BIOCHEMICAL OXYGEN DEMAND CURVE), gas dissolution into water, and decay of radioactive elements.

Formulas

In first-order kinetics, the rate of decomposition or decay is proportional to the concentration of remaining material (chemicals or microorganisms):

$$\frac{-d[\mathrm{C}]}{dt} = k[\mathrm{C}]$$

Integration of the equation above gives

$$[C_t] = [C_0]e^{-kt}$$

where

$[C_0]$ = initial concentration (mol/L or number of microorganisms/L)

$[C_t]$ = chemical or microbial concentration (mol/L or number of microorganisms/L) at time t

k = decay rate (1/time)

A useful form of the first-order equation is

$$\ln\frac{[C_0]}{[C_t]} = kt$$

Using the $\log_{10}$ basis, the equation above is written as

$$\log_{10}\frac{[C_0]}{[C_t]} = \frac{kt}{2.303} = k't \qquad k = 2.303k'$$

A plot of $\log_{10}$ $[C_0]/[C_t]$ versus t gives a straight line with a slope of k' (see Figure C1).

The time at which $[C_t] = \frac{1}{2}[C_0]$ is called the *half-life* $t_{1/2}$:

$$t_{1/2} = \frac{0.693}{k}$$

Reference

Sawyer, C. N. and P. L. McCarty. 1978. *Chemistry for Environmental Engineering*, McGraw-Hill, New York.

RECIRCULATION RATIO: Trickling Filters

Introduction

Trickling filter effluents are sometimes partially recirculated through the filter to increase the treatment efficiency of the filter medium (Metcalf and Eddy, 1991). A portion of the treated effluent is returned to the filter. The recirculation ratio R is the ratio of the flow rate of recirculated effluent to the flow rate of the wastewater influent (Nathanson, 1986):

Formula

$$R = \frac{Q_R}{Q}$$

where

Q_R = flow rate of recirculated trickling filter effluent (L/min)
Q = flow rate of wastewater influent (L/min)

References

Metcalf and Eddy, Inc. 1991. *Wastewater Engineering: Treatment, Disposal and Reuse.* 3rd ed., McGraw-Hill, New York.

Nathanson, J. A. 1986. *Basic Environmental Technology: Water Supply, Waste Disposal and Pollution Control*, Wiley, New York.

REDOX POTENTIAL: Oxidation–Reduction Potential

See also NERNST EQUATION.

Definition

The oxidation-reduction potential or *redox potential* E_h of an aquatic system is an index of the ability to donate (oxidation) or accept (reduction) electrons. This index is closely related to biological processes. (Raymont, 1980).

Formula

The redox condition for equilibrium processes can be expressed as pE (dimensionless) or as E_h (V). pE is a measure of electron activity in solution.

PE and E_h are related by the following equation (Eriksson, 1985; Freeze and Cherry, 1979; Latimer, 1956):

$$E_h = \frac{2.3RT}{nF} \mathrm{pE}$$

where

R = gas constant (8.31450 J/mol · K)
T = absolute temperature (K)
n = number of electrons participating in the reaction
F = Faraday constant (9.65×10^4 C/mol)

For reactions involving the transfer of a single electron at 25°C, the expression above becomes

$$E_h = 0.059 \mathrm{pE}$$

References

Eriksson, E. 1985. *Principles and Applications of Hydrochemistry*, Chapman & Hall, London.

Freeze, R. A., and J. A. Cherry. 1979. *Groundwater*, Prentice Hall, Upper Saddle River, NJ.

Latimer, W. M. 1956. *Oxidation Potentials*, Prentice Hall, Upper Saddle River, NJ.

Raymont, J. E. G. 1980. *Plankton and Productivity in the Oceans*, vol. 1, *Phytoplankton*, 2nd ed., Pergamon Press, Oxford.

Stumm, W., and J. J. Morgan. 1981. *Aquatic Chemistry*, Wiley-Interscience, New York.

RESPIRATORY QUOTIENT

Formula

The respiratory quotient RQ is given by the following:

$$\mathrm{RQ} = \frac{Q_{\mathrm{CO2}}}{Q_{\mathrm{O2}}}$$

where

Q_{CO2} = rate of carbon dioxide formation (mol/L · h)
Q_{O2} = rate of oxygen uptake (mol/L · h)

Reference

Sukatsch, D. A., and A. Dziengel. 1987. *Biotechnology: A Handbook of Practical Formulae*, Longman Scientific & Technical, London.

RETARDATION FACTOR: Groundwater Contaminants

Introduction

The partition or sorption of contaminants on soils or aquifer materials retards their movement compared to the movement of water. Contaminant retardation is expressed by a retardation factor, which is related to the solid/water distribution coefficient K_d by the following formula (Roberts et al., 1980; Sawyer et al., 1994):

Formula

$$t_r = 1 + \rho K_d + e$$

where

t_r = retardation factor (or ratio of water movement rate to contaminant movement rate)

ρ = bulk density of soil (kg/L)

K_d = distribution coefficient (L/kg)

e = void fraction of soil

References

Roberts, P. V., M. Reinhard, and P. L. McCarty. 1980. Organic contaminant behavior during groundwater recharge. J. Water Pollut. Control Fed. 52: 161–172.

Sawyer, C. N., P. L. McCarty, and G. F. Parkin. 1994. *Chemistry for Environmental Engineering*, McGraw-Hill, New York.

REYNOLDS NUMBER

Introduction

The Reynolds number Re describes the nature of flow (i.e., whether laminar or turbulent) in a pipe, channel, or lake. In turbulent flow, the inertial effects are of greater importance than the viscous effects that characterize laminar flow. The

Reynolds number may be viewed as a ratio of shear stress due to turbulence to shear stress due to viscosity.

Formula

$$\mathrm{Re} = \frac{Ud}{\mu}$$

where

U = water velocity or flow velocity (cm/s)

d = characteristic length (cross section) (for a lake it is the depth of thickness of water layer) (cm)

μ = kinematic viscosity of water (cm^2/s)

Numerical Values

In a pipe, a Reynolds number above approximately 2000 indicates turbulent flow. Turbulence is well established at $\mathrm{Re} > 4000$ (Thibodeaux, 1979). For a typical lake, the flow is turbulent, and $U = 10\,\mathrm{cm/s}$, $d = 10\,\mathrm{m}$, $\mu = 0.01\,\mathrm{cm^2/s}$, leading to $\mathrm{Re} = 10^6$ (Goldman and Horne, 1983). Most natural streams display turbulent flow.

References

Crueger, W., and A. Crueger. 1989. *Biotechnology: A Textbook of Industrial Microbiology*, Sinauer Associates, Sunderland, MA.

Goldman, C. R., and A. J. Horne. 1983. *Limnology*, McGraw-Hill, New York.

Streeter, V. L. 1971. *Fluid Mechanics*, 5th ed., McGraw-Hill, New York.

Thibodeaux, L. J. 1979. *Chemodynamics: Environmental Movement of Chemicals in Air, Water, and Soil*, Wiley. New York.

RNA CONTENT OF BACTERIA

Introduction

RNA content of cells gives an indication of cell activity. The growth rate of growing cells was found to be correlated with RNA content of cells (Båmstedt and Skjoldal, 1980; Dortch et al., 1983; Kemp et al., 1993; Poulsen et al., 1993). RNA content of cells can be determined by ethidium bromide fluorometry and can also be estimated by using 16S rRNA probes (DeLong et al., 1989; Lee and Kemp, 1994; Lee et al., 1993).

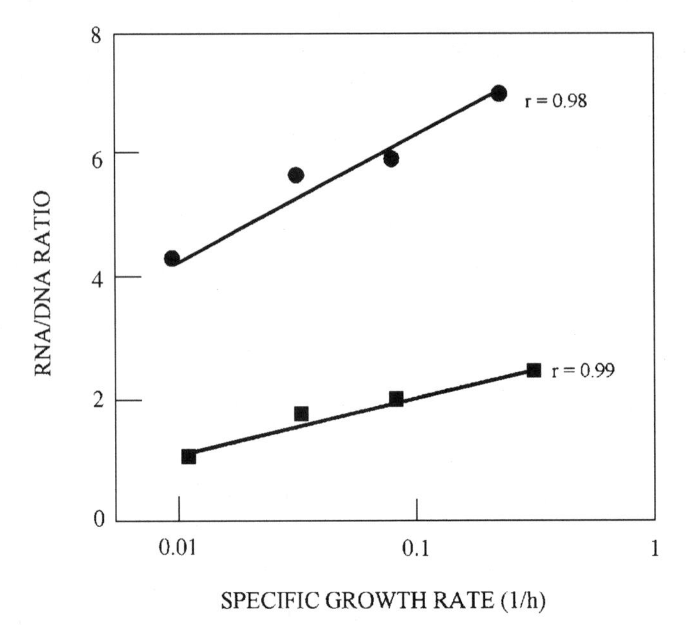

FIGURE R1: Relationship Between the Specific Growth Rate μ and the Cell RNA/DNA Ratio (*Source:* Adapted from Kemp et al., 1993.)

Numerical Values

- *Natural marine bacterioplankton*

  RNA content = 1.6–5.4 fg/cell (ethidium bromide fluorometry) (Lee and Kemp, 1994).

  RNA content = 1.9–9.5 fg/cell (planktonic marine bacteria; indirect calculation from cell size; Simon and Azam, 1989).
- *Marine bacterial isolates:* Depending on the growth rate and type of bacteria, RNA content varied from approximately <10 to 55 fg/cell.

The *RNA/DNA ratio* was found to be strongly correlated with the specific growth rate of bacteria (see Figure R1).

References

Båmstedt, U., and H. R. Skjoldal. 1980. RNA concentration of zooplankton: relationship with size and growth. Limnol. Oceanogr. 25: 304–316.

DeLong, E. F., G. S. Wickham, and N. R. Pace. 1989. Phylogenetic stains: ribosomal RNA-based probes for the identification of single cells. Science 243: 1360–1363.

Dortch, Q., T. L. Roberts, J. R. Clayton, Jr., and S. I. Ahmed. 1983. RNA/DNA ratios and DNA concentrations as indicators of growth rate and biomass in planktonic marine organisms. Mar. Ecol. Prog. Ser. 13: 61–71.

Kemp, P. E., S. Lee, and J. LaRoche. 1993. Estimating the growth rate of slowly growing marine bacteria from RNA content. Appl. Environ. Microbiol. 59: 2594–2601.

Lee, S. and P. E. Kemp. 1994. Single-cell RNA content of natural planktonic bacteria measured by hybridization with multiple 16S rRNA-targeted fluoresence probes. Limnol. Oceanogr. 39: 869–879.

Lee, S. C., Malone, and P. E. Kemp. 1993. Use of multiple 16S rRNA-targeted fluorescent probes to increase signal strength and measure cellular RNA from natural planktonic bacteria. Mar. Ecol. Prog. Ser. 101: 193–201.

Poulsen, L. K., G. Ballard, and D. A. Stahl. 1993. Use of rRNA fluorescence *in situ* hybridization for measuring the activity of single cells in young and established biofilms. Appl. Environ. Microbiol. 59: 1354–1360.

Simon, M., and F. Azam. 1989. Protein content and protein synthesis rates of planktonic marine bacteria. Mar. Ecol. Prog. Ser. 51: 201–213.

S

SAND FILTRATION

See FILTER COEFFICIENT.

SHANNON–WEAVER INDEX

See DIVERSITY INDEX.

SHORELINE DEVELOPMENT INDEX

Definition

The shoreline development index relates the shoreline length of a lake to the circumference of a circle with the same surface area (Cole, 1979).

Formula

$$D_L = \frac{L}{2\sqrt{\pi A}}$$

where

L = lake shoreline length (m or km)
A = lake surface area (m^2 or km^2)

Numerical Values

A nearly circular lake has an index close to 1. Some elongated lakes have $D_L = 3.3$–3.4 (Cole, 1979).

Reference

Cole, G. A. 1979. *Textbook of Limnology*, C.V. Mosby, St. Louis, MO.

SIMILARITY COEFFICIENT

Definition

The similarity coefficient expresses the degree of relatedness between microbial strains. The coefficient S, expressing the similarity between two stains A and B, is the ratio of similar characters to the total number of characters under examination.

Formula

$$S = \frac{a + d}{a + b + c + d} \qquad 0 < S < 1$$

where

$a + d =$ the sum of characters common to strains A and B (e.g., a: both strains are positive for characters a and negative for characters d)

$b + c =$ sums of characters not common to both strains A and B (e.g., b: characters for which A is positive and B is negative; c: characters for which A is negative and B positive)

Reference

Schlegel, H. G. 1985. *General Microbiology*, Cambridge University Press, Cambridge.

SIMILARITY INDEX

Formula

The similarity index S between two samples is given by the following:

$$S = \frac{2C}{A + B}$$

where

$A =$ number of species in sample A

$B =$ number of species in sample B

$C =$ numbers of species common to both samples A and B

Reference

Jorgensen, S. E., and I. Johnsen. 1989. *Principles of Environmental Science and Technology*, Elsevier, Amsterdam.

SIMPSON INDEX

See DIVERSITY INDEX.

SLUDGE AGE

Definition/Introduction

Sludge age is the mean cell residence time (MCRT) in a bioreactor. While the hydraulic retention time may be on the order of hours, the mean cell residence time may be on the order of days. This parameter is the reciprocal of the microbial specific growth rate μ.

Formula

Sludge age θ_c is given by the following formula (Hammer, 1986; Curds and Hawkes, 1983):

$$\theta_c = \frac{\text{MLSS} \times V}{\text{SS}_e \times Q_e + \text{SS}_w \times Q_w}$$

where

θ_c = sludge age (days)
MLSS = mixed liquor suspended solids (mg/L)
V = volume of aeration tank (m^3)
SS_e = suspended soils in wastewater effluent (mg/L)
Q_e = quantity of wastewater effluent (m^3/day)
SS_w = suspended solids in wasted sludge (mg/L)
Q_w = quantity of wasted sludge (m^3/day)

Numerical Values

Sludge age may vary from 3 to 14 days in conventional activated sludge. It may be lower than 3 days in high-rate activated sludge and 5–25 days in nitrifying activated sludge (Sundstrom and Klei, 1979).

References

Curds, C. R., and H. A. Hawkes, Eds. 1983. *Ecological Aspects of Used-Water Treatment*, vol. 2, Academic Press, London.

Hammer, M. J. 1986. *Water and Wastewater Technology*, Wiley, New York.

Sundstrom, D. W., and H. E. Klei. 1979. *Wastewater Treatment*, Prentice Hall, Upper Saddle River, NJ.

SLUDGE DENSITY INDEX (SDI)

See also SLUDGE VOLUME INDEX.

Definition

Sludge density index is the density of settling sludge after a 30-minute settling period. Sludge density index is essentially the reciprocal of the sludge volume index (SVI).

Formula

$$\mathrm{SDI} = \frac{\mathrm{MLSS}}{V}$$

where

SDI = sludge density index (g/mL)

MLSS = mixed liquor suspended solids concentration (g/L)

V = settled sludge after 30 minute settling time (mL/L)

Numerical Values

SDI is expressed in g/mL and is thus the mass of sludge per specified sludge volume. The SDI varies from about 0.02 g/mL for a sludge that settles well to about 0.005 g/mL or less for a bulking sludge.

Reference

Institute of Water Pollution Control, 1987. *Unit Processes: Activated Sludge*, IWPC, Maidstone, Kent, England.

SLUDGE FUEL VALUE

Definition

Fuel value is the calorific potential of wastewater sludges for sustaining combustion. It is also called the *heat value* of sludge.

Formula

The following empirical formula was proposed by Fair et al. (1968):

$$Q_H = A\frac{100P_v}{100 - P_c} - B\frac{100 - P_c}{100}$$

where

Q_H = heat value of sludge (Btu/lb dry solids)
P_v = % volatile solids
P_c = dosage of conditioning chemicals used for sludge dewatering (% weight dry solids)
A = empirical constant ($A = 107$ for activated sludge)
B = empirical constant ($B = 5$ for activated sludge)

Numerical Values

A value of 10,000 Btu/lb dry volatile solids is generally used. In comparison, coal has a fuel value of 14,000 Btu/lb and petroleum has a value of 20,000 Btu/lb (Weber, 1972).

References

Fair, G. W., J. C. Geyer, and D. A. Okun. 1968. *Water and Wastewater Engineering*, vol. II, Wiley, New York.

Weber, W. J., Jr., Ed. 1972. *Physicochemical Processes for Water Quality Control*, Wiley-Interscience, New York.

SLUDGE QUANTITIES: Wastewater Treatment Processes

Numerical Values

The typical sludge quantities generated by three domestic wastewater treatment processes are given in Table S1.

TABLE S1: Typical Sludge Quantities Generated by Domestic Wastewater Treatment Plants

Treatment Process	Volume (m^3 sludge per 1000 m^3 wastewater)	Mass (kg solid per m^3 wastewater)
Primary settling	3.0	0.144
Trickling filter	0.7	0.054
Activated sludge	19	0.216

Source: Adapted from Sundstrom and Klei (1979).

Reference

Sundstrom, D. W., and H. E. Klei. 1979. *Wastewater Treatment*, Prentice Hall, Upper Saddle River, NJ.

SLUDGE VOLUME INDEX (SVI)

See also DILUTED SLUDGE VOLUME INDEX (DSVI); STIRRED SPECIFIC VOLUME INDEX (SSVI); SLUDGE DENSITY INDEX.

Definition

The sludge volume index (SVI) is the volume occupied by 1 g of sludge after a 30-minute settling. It is used as an index for assessing the settleability of activated sludge or other suspensions. It is well known that overgrowth of filamentous bacteria in activated sludge leads to an increase in sludge volume index. This phenomenon is called *filamentous bulking*.

Formula

$$\mathrm{SVI} = \frac{V}{V_0 \times \mathrm{MLSS}}$$

where

SVI = sludge volume index (mL/g)
V = settled sludge volume after a 30-minute settling (mL)
V_0 = initial volume of sludge tested (L)
MLSS = mixed liquor suspended solids (g/L)

Numerical Values

An acceptable range for sludge SVI is between 35 and 100 mL/g (Sundstrom and Klei, 1979).

Some have proposed the use of stirred sludge volume index (SSVI), which is carried out under stirred conditions (*see* STIRRED SLUDGE VOLUME INDEX).

References

Dick, R. I., and P. A. Vesilind. 1969. The SVI—what is it? J. Water Polut. Control Fed. 41: 1285–1289.

Donaldson, W. 1932. Some notes on the question of sewage treatment works. Sewage Works J. 4: 48–52.

Sundstrom, D. W., and H. E. Klei. 1979. *Wastewater Treatment*, Prentice Hall, Upper Saddle River, NJ.

SMOLUCHOWSKI EQUATION

Also called the *Helmotz–Smoluchowski equation*.

Introduction

The Smoluchowski equation gives the relationship between zeta potential and electrophoretic mobility of colloidal particles, including biocolloids.

Formula

$$U = \frac{D\zeta}{4\pi\eta}$$

where

U = electrophoretic mobility (μm/s · V · cm)

D = dielectric constant

ζ = zeta potential (V)

η = viscosity of suspending fluid

References

Marshall, K. C. 1976. *Interfaces in Microbial Ecology*, Harvard University Press, Cambridge, MA.

Tadros, T. F. 1980. Particle-surface adhesion, pp. 93–116. In: *Microbial Adhesion to Surfaces*, R. C. W. Berkeley, J. M. Lynch, J. Melling, P. R. Rutter, and B. Vincent, Eds., Ellis Horwood, Chichester, West Sussex, England.

SODIUM ADSORPTION RATIO (SAR)

Introduction

Sodium ions alter soil permeability. The sodium adsorption ratio indicates whether or not the sodium content of a wastewater is high enough to cause infiltration problems in soils.

Formula

$$SAR = \frac{[Na^+]}{(0.5[Ca^{2+} + Mg^{2+}])^{0.5}}$$

$[Na^+]$, $[Ca^{2+}]$, and $[Mg^{2+}]$ are expressed in milliequivalents/L.

The higher the SAR, the higher the tendency of sodium to absorb to the cation exchange sites (Donahue et al., 1977). Soil permeability is affected when $SAR > 9$ (Rich, 1980). High sodium levels are also toxic to plants.

References

Ayers, R. S., and K. K. Tanjy. 1981. Agronomic aspects of crop irrigation with wastewater, p. 578. In: *Water Forum '81*, vol. 1, American Society of Civil Engineers, New York.

Donahue, R. L., R. W. Miller, and J. C. Shikluma. 1977. *Soils: An Introduction to Soils and Plant Growth*, Prentice Hall, Upper Saddle River, NJ.

Rich, L. G. 1980. *Low-Maintenance, Mechanically Simple Wastewater Treatment Systems*, McGraw-Hill, New York.

SOIL

See BULK DENSITY OF SOILS; PARTITION COEFFICIENTS (the various subentries); POROSITY, SOIL; VOID RATIO, SOIL; WATER CONTENT, SOIL; WATER POTENTIAL, SOIL.

SOIL: Volatilization of Organic Chemicals from Soil Surface

Introduction

Some organic chemicals in soil undergo volatilization into the atmosphere. The rate of volatilization of a given chemical depends on several factors, which include characteristics of the chemicals (e.g., vapor pressure of the chemical, adsorption to soil, concentration with depth) as well as soil characteristics (e.g., porosity, texture, bulk density, water and organic matter contents, air temperature) (Donnely et al., 1994; Spencer et al., 1982).

Formula

The vapor flux from the soil surface into the atmosphere is given by (Spencer, 1982)

$$J = \frac{D_o P_a^{10/3}(C_2 - C_s)}{P_T L}$$

where

J = vapor flux from soil surface ($\mu g/cm^2 \cdot day$)
D_o = vapor diffusion coefficient in air ($cm^2 \cdot day$)
P_a = soil air-filled porosity (cm^3/cm^3)
C_2 = air concentration of chemical at the soil surface ($\mu g/L$)
C_s = air concentration of chemical at the bottom of soil layer ($\mu g/L$)
P_T = total soil porosity
L = soil depth (cm)

References

Donnely, K. C., C. S. Anderson, G. C. Barbee, and D. J. Manek. 1994. Soil toxicology, pp. 321–352. In: *Basic Environmental Toxicology*, L. G. Cockerham and B. S. Shane, Eds., CRC Press, Boca Raton, FL.

Spencer, W. F., W. J. Farmer, and W. A. Jury. 1982. Behavior of organic chemicals at soil, air, water interfaces as related to predicting the transport and volatilization of organic pollutants. Environ. Toxicol. Chem. 1: 17–26.

SOIL ORGANIC MATTER/SOIL WATER DISTRIBUTION COEFFICIENT (K_{oc})

See PARTITION COEFFICIENTS: Soil Organic Matter/Soil Water.

SOIL WATER/AIR DISTRIBUTION COEFFICIENT

See PARTITION COEFFICIENT: Soil Water/Air

SOLAR IRRADIANCE

Definition

Solar irradiance is the quantity of energy received on a unit area of a body of water over time.

Formula

In an aquatic environment (e.g., lake), the solar irradiance at depth z is given by

$$I_z = I_0 e^{-\eta z}$$

where

I_z = solar irradiance at depth z ($J/m^2 \cdot s$ or $W/m^2 \cdot s$ or $gcal/cm^2 \cdot min$)
I_0 = solar irradiance at the water surface ($J/m^2 \cdot s$ or $W/m^2 \cdot s$ or $gcal/cm^2 \cdot min$)
η = extinction coefficient
z = depth (m)

Reference

Wetzel, R. G., and G. E. Likens. 1991. *Limnological Analyses*, 2nd ed., Springer-Verlag, New York.

SOLID/WATER DISTRIBUTION COEFFICIENT

See PARTITION COEFFICIENT: Solid/Water.

SPECIES RICHNESS

See DIVERSITY INDEX.

TABLE S2: Specific Gravity of Selected Solids

Solid	Specific Gravity
Clay	1.28
Diatomaceous earth	0.22
Glass, silica	2.72
Ice at 0°C	0.90
Earth crust	2.67

Source: Adapted from Thibodeaux (1979).

SPECIFIC GRAVITY

Definition

The specific gravity of a substance is the ratio of its density to the density of water at 4°C.

Formula

$$\text{Specific gravity} = \frac{\rho_{\text{substance}}}{\rho_{\text{water}}}$$

where

$\rho_{\text{substance}}$ = density of substance (g/cm^3)
ρ_{water} = density of water at 4°C = 1 g/cm^3

Specific gravity is a dimensionless number.

Numerical Values

See Table S2.

References

Peters, E. I. 1976. *Problem Solving for Chemistry*, Saunders College Publishing, Philadelphia.

Thibodeaux, L. J. 1979. *Chemodynamics*, Wiley, New York.

SPECIFIC GROWTH RATE: Microorganisms

See MONOD's EQUATION.

SPECIFIC GROWTH RATE: Relation with Doubling Time of Microorganisms

Introduction

The specific growth rate μ is inversely proportional to a microbial culture doubling time.

Formula

$$\mu = \frac{\ln 2}{t_d} = \frac{0.693}{t_d}$$

where

t_d = microbial culture doubling time (h)
μ = specific growth rate (h^{-1})

Reference

Slater, J. H. 1979. Microbial population and community dynamics, pp. 45–63. In: *Microbial Ecology: A Conceptual Approach*, J. M. Lynch and N. J. Poole, Eds., Blackwell Scientific, Oxford.

SPECTROPHOTOMETER

See BEER–LAMBERT LAW.

STANDARD FREE ENERGY

See also FREE ENERGY.

Definition

The standard free energy change is the change in free energy for any chemical reaction occuring under standard conditions of temperature, pressure, and concentration.

Formulas

Consider the following chemical reaction (Sawyer et al., 1994):

$$a\mathrm{A} + b\mathrm{B} \rightleftharpoons c\mathrm{C} + d\mathrm{D}$$

The free energy of the reaction above is given by

$$\Delta G = \Delta G^0 + RT \ln \frac{[C]^c[D]^d}{[A]^a[B]^b}$$

where

ΔG = free energy (J)
ΔG^0 = standard free energy (J)
R = universal gas constant = 8.314 J/K · mol
A = absolute temperature (K)

The reaction will proceed to the right is ΔG is negative, and to the left if ΔG is positive. At equilibrium, ΔG is equal to zero, and the standard free energy ΔG^0 is

$$\Delta G^0 = -RT \ln \frac{[C]^c[D]^d}{[A]^a[B]^b}$$
$$= -RT \ln K'_{eq}$$

where

$$K'_{eq} = \text{equilibrium constant} = \frac{[C]^c[D]^d}{[A]^a[B]^b}$$

Numerical Values

See Table S3.

TABLE S3: Relationship Between the Equilibrium Constant K'_{eq} and Standard Free Energy Change at 25°C

K'_{eq}	ΔG^o (cal)
0.001	+4089
0.01	+2726
0.1	+1363
1.0	0
10.0	−1363
100.0	−2726
1000.0	−4089

Source: Adapted from Lehninger (1973).

References

Lehninger, A. L. 1973. *Short Course in Biochemistry*, Worth Publishers, New York.

Sawyer, C. N., P. L. McCarty, and G. F. Parkin. 1994. *Chemistry for Environmental Engineering*, McGraw-Hill, New York.

STANDARD OXIDATION–REDUCTION POTENTIAL (E'_0)

See also NERNST EQUATION.

Definition/Introduction

The standard oxidation–reduction potential (E'_0) is the electromotive force (in volts) given by an electrode immersed in a solution containing 1 *M* concentration of both an electron donor and its conjugate electron acceptor at 25°C and pH 7.0. The electrode (*half-cell*) is connected to a reference half-cell (hydrogen electrode) which has an electromotive force of − 0.41 V at 25°C and pH 7.0. The more negative the potential, the higher the tendency to lose electrons. Conversely, the more positive the potential, the higher the tendency to accept electrons (Table S4).

TABLE S4: Standard Oxidoreduction Potentials E'_0 for Some Components of the Electron Transport Chain

Redox Couple	E'_0 (V)
Substrate couples	
Acetyl-CoA + CO_2 + $2H^+$ + $2e^-$ → pyruvate + CoA	−0.48
Pyruvate + 2H + $2e^-$ → lactate	−0.19
Fumarate + $2H^+$ + $2e^-$ → succinate	+0.03
Electron transport chain	
$2H^+ + 2e^- \rightarrow H_2$	−0.41
$NAD^+ + H^+ + 2e^- \rightarrow NADH$	−0.32
Ubiquinone + $2H^+$ + $2e^-$ → ubiquinol	+0.04
Cytochrome b (ox) + e^- → cytochrome b (red)	+0.07
Cytochrome c_1 (ox) + e^- → cytochrome c_1 (red)	+0.23
Cytochrome c (ox) + e^- → cytochrome c (red)	+0.25
Cytochrome a (ox) + e^- → cytochrome a (red)	+0.29
Cytochrome a_3 (ox) + e^- → cytochrome a_3 (red)	+0.55
$1/2O_2 + 2H^+ + 2e^- \rightarrow H_2O$	+0.82

Source: Adapted from Lehninger (1982).

Formula

Electrons tend to flow from an electronegative redox couple toward a more electropositive one. As they flow down the electron transport chain, they lose free energy. The free-energy change is given by the following formula:

$$\Delta G^0 = -nF\Delta E_0'$$

where

ΔG^0 = standard free energy change (cal)

n = number of electrons transferred

F = the Faraday constant (23,062 cal/V · mol)

$\Delta E_0'$ = difference between the standard reduction potential of the donor system and that of the acceptor system (V)

Reference

Lehninger, A. L. 1982. *Principles of Biochemistry*, Worth Publishers, New York.

STEFAN–BOLTZMANN EQUATION

Introduction

The Stefan–Boltzmann equation gives the amount of energy radiated by a body (e.g., sun) as a function of its temperature.

Formula

$$\text{Energy radiated/m}^2 \cdot \text{s} = \sigma T^4$$

where

σ = Stefan–Boltzmann constant = $5.67 \times 10^{-8}\ \text{J/m}^2 \cdot \text{s} \cdot \text{K}^4$

T = absolute temperature (K)

Reference

Tinoco, I., Jr., K. Sauer, and J. C. Wang. 1995. *Physical Chemistry: Principles and Applications in Biological Sciences*, Prentice Hall, Upper Saddle River, NJ.

STEFAN-BOLTZMANN LAW

Introduction

The Stefan–Boltzmann law pertains to the increase of equilibrium temperature on Earth as a result of anthropogenic input of energy.

Formula

$$\frac{Q+q}{Q} = \left(T + \frac{t}{T}\right)^4$$

If t is much lower than T, and q much lower than Q, the equation above becomes

$$t = \frac{T}{4Q}q$$

where

t = increase in equilibrium temperature on earth as a result of anthropogenic input of energy (°C)

T = equilibrium temperature of the earth's surface (°C)

Q = annual supply of heat from solar radiation

q = annual supply of heat from fossil fuels

Reference

Ramade, F. 1981. *Ecology of Natural Resources*, Wiley, New York.

STIRRED SPECIFIC VOLUME INDEX (SSVI)

Introduction

The measurement of stirred specific volume index (SSVI) is carried out using a cylinder of specified dimensions that is fitted with a stirrer rotating at one revolution per minute. The stirred specific volume index (SSVI) is usually carried at an MLSS (mixed liquor suspended solids) concentration of 3.5 g/L, due to the effect of suspended solid concentration on sludge settleability ($SSVI_{3.5}$) (Institute of Water Pollution Control, 1987; Water Research Centre, 1975).

Formula

$$\mathrm{SSVI} = \frac{V}{\mathrm{MLSS}}$$

where

V = settled volume after 30 minute (ml)

MLSS = mixed liquor suspended solids (g/L)

Numerical Values

- Good settling sludge: SSVI = 40–50 mL/g
- Poor settling sludge: SSVI > 120 mL/g

References

Institute of Water Pollution Control. 1987. *Unit Processes: Activated Sludge*, IWPC, Maidstone, Kent, England.

Water Research Centre. 1975. *Settling of Activated Sludge*, Technical Report TR11, WRC, Stevenage, Hertfordshire, England.

STOKES–EINSTEIN EQUATION

Introduction

The Stokes–Einstein equation gives the diffusion coefficient D of colloidal particles, including biocolloids (e.g., viruses).

Formula

$$D = \frac{kT}{6\pi\eta r} = \frac{RT}{6\pi\eta r N}$$

where

k = Boltzmann constant ($k = 1.38 \times 10^{-23}$ J/K)

T = absolute temperature (K)

R = gas constant = 8.314510 J/mol · K

η = viscosity of the liquid ($\mathrm{N/cm^2 \cdot s}$)

r = particle radius

N = Avogadro number = $6.022 \times 10^{23}\ \mathrm{mol^{-1}}$

References

Stumm, W., and J. J. Morgan. 1981. *Aquatic Chemistry*, Wiley-Interscience, New York.

Tadros, T. F. 1980. Particle–surface adhesion, pp. 93–116. In: *Microbial Adhesion to Surfaces*, R. C. W. Berkeley, J. M. Lynch, J. Melling, P. R. Rutter, and B. Vincent, Eds., Ellis Horwood, Chichester, West Sussex, England.

STOKES' LAW

Introduction

Stokes' law deals with particle settling, and states that the settling velocity of a particle is proportional to the square of the particle diameter. Stokes' law applies to smooth rigid spherical bodies sinking in a fluid under laminar flow conditions.

Formula

$$V_c = \frac{g(p_s - p)d^2}{18\eta}$$

where

V_c = terminal sinking velocity (m/s)

g = gravity constant (gravitational acceleration) (m/s^2) = 9.80 m/s^2

p_s = density of particle (kg/m^3)

p = density of fluid (kg/m^3)

d = particle diameter (m)

η = viscosity of suspending fluid ($kg/m \cdot s$)

References

Metcalf and Eddy, Inc. 1991. *Wastewater Engineering: Treatment, Disposal and Reuse*, 3rd ed., McGraw-Hill, New York.

Schwarzenbach, R. P., P. M. Gschwend, and D. M. Imboden. 1993. *Environmental Organic Chemistry*, Wiley-Interscience, New York.

STREETER–PHELPS EQUATION

Also called the dissolved oxygen sag equation.

Introduction

When a wastewater treatment plant effluent is discharged into a stream, the dissolved oxygen in the stream decreases to a minimum, then gradually increases at increasing distance from the discharge point. Figure S1 shows the plot of dissolved oxygen concentration as a function of the distance from the discharge point. This plot is called the *DO sag curve.*

Two competing processes that affect the level of dissolved oxygen in the receiving stream are:

- *Reaeration* is the entry of oxygen from the atmosphere into the water.
- Biochemical oxygen demand or deoxygenation is exerted by the stream organisms.

Formulas

The DO sag equation is as follows:

$$D = \frac{k_1 L_0}{k_2 - k_1}(e^{-k_1 t} - e^{-k_2 t}) + D_0 e^{-k_2 t}$$

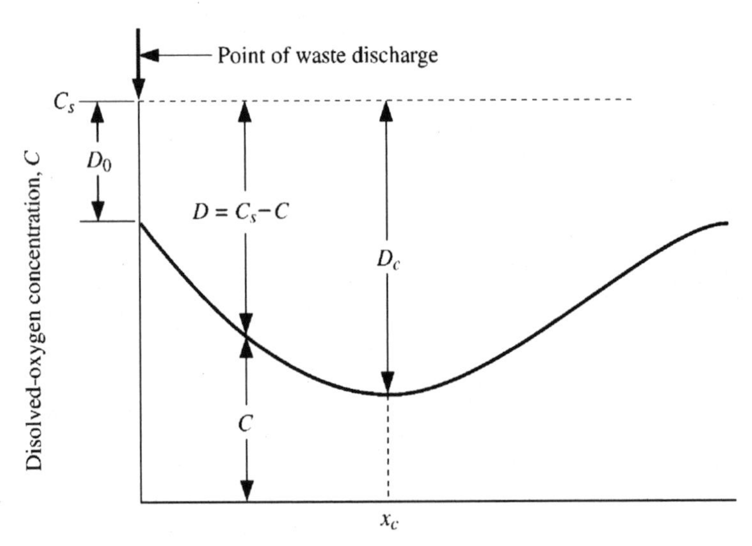

FIGURE S1: Typical Dissolved Oxygen Sag Curve (*Source:* Metcalf and Eddy, 1991.)

where

$\frac{k_1 L_0}{k_2 - k_1}(e^{-k_1 t} - e^{-k_2 t})$ = BOD exertion

$D_0 e^{-k_2 t}$ = initial deficit

D = dissolved oxygen (DO) deficit (mg/L)

= $C_s - C$ (C_s = saturation DO concentration; C = actual DO concentration of the stream water in mg/L)

k_1 = deoxygenation constant (BOD decay coefficient) (day^{-1})

k_2 = reaeration constant (day^{-1})

L_0 = initial BOD at time $t = 0$ (mg/L)

t = travel time = x/U (U = stream velocity)

D_0 = DO deficit (mg/L) at distance $x = 0$

t_{max}, the time to reach the maximum deficit D_c or the minimum DO, is given by the equation

$$t_{max} = \frac{1}{k_2 - k_1} \ln\left[\frac{k_2}{k_1}\left(1 - \frac{D_0(k_2 - k_1)}{k_1 L_0}\right)\right]$$

$$D_c = \frac{k_1}{k_2} L_0 e^{-k_1 t_{max}}$$

References

Metcalf and Eddy, Inc. 1991. *Wastewater Engineering: Treatment, Disposal and Reuse*, 3rd ed., McGraw-Hill, New York.

Ray, B. T. 1995. *Environmental Engineering*, PWS, Boston.

Streeter, H. W., and E. B. Phelps. 1925. A study of the pollution and natural purification of the Ohio River, Public Health Bulletin 146, U.S. Public Health Service, Washington, DC.

SUBSTRATE UTILIZATION RATE OF MICROORGANISMS

Definition

In a biological reactor, the substrate utilization (removal) rate is the mass of substrate removed per mass of microorganisms per time.

Formulas

$$q = \frac{\text{mass substrate removed}}{\text{mass microorganisms.time}} = \frac{S_0 - S_t}{Xt}$$

where

S_0 = initial substrate concentration (mg/L)
S_t = substrate concentration at time t (mg/L)
X = microorganism concentration (mg/L)
t = time

The relationship of q with the specific growth rate μ and the growth yield Y is given by

$$q = \frac{\mu}{Y}$$

where

μ = specific growth rate (1/time)
Y = growth yield

Reference

Vesilind, P. A., and J. J. Pierce. 1982. *Environmental Engineering*, Ann Arbor Science, Ann Arbor, MI.

SURFACE: pH at the Surface of Charged Particles

Introduction

The pH at the surface of charged particles may be different from that in the bulk phase. There is a relationship between the pH at the surface (pH_s) and the pH in the bulk phase (pH_b). pH_s is generally lower than pH_b. This difference in pH affects enzyme and microbial activity at interfaces.

Formula

$$pH_s = pH_b + 0.325u$$

where

pH_s = pH at the surface of charged particle

pH_b = pH of the bulk surrounding phase

u = electrophoretic mobility of the charged particle (μm/s · V · cm)

Reference

McLaren, A. D., and J. Skujins. 1968. The physical environment of microorganisms in soil, pp. 3–24. In: *The Ecology of Soil Bacteria*, T. R. G. Gray and D. Parkinson, Eds., Liverpool University Press, Liverpool, Lancashire, England.

SURFACE MICROBIAL COLONIZATION EQUATION

Introduction

Several factors affect the colonization of a surface by bacteria. These factors include attachment, detachment, predation, growth, death, and interaction between colonies. Surface colonization can be simplified by taking into account only the attachment of microorganisms to a given surface and their growth on the surface.

Formulas

The following equation describes the colonization rate, which is equal to the sum of the attachment rate (A) and the growth rate (μN) (Brannan and Caldwell, 1986; Caldwell et al., 1981):

$$\frac{dN}{dt} = \mu N + A$$

Integration gives the following equation, which predicts the number of cells N at the surface at any time t:

$$N = \frac{A}{\mu} e^{\mu t} - \frac{A}{\mu}$$

where

N = number of microorganisms on the surface

A = attachment rate (cells/h)

μ = specific growth rate (h^{-1})

t = incubation time (h)

References

Brannan, D. K., and D. E. Caldwell. 1986. Ecology and metabolism of *Thermothrix thiopara*. Adv. Appl. Microbiol. 31: 233–270.

Caldwell, D. E., D. K. Brannan, M. E. Morris, and M. R. Betlach. 1981. Quantitation of microbial growth on surfaces. Microb. Ecol. 7: 1–11.

SURFACE TENSION

Definition

Surface tension is the free energy per unit surface area or the force per unit length on a surface. Surfactants lower the surface tension and, thus, the free energy at the surface.

Formula

The surface tension is given by

$$G_{\text{surface}} = \gamma A$$

where

G_{surface} = surface energy (mJ)
γ = surface tension (mJ/m^2 or mN/m)
A = surface area (m^2)

Numerical Values

See Table S5.

References

Tinoco, I., Jr., K. Sauer, and J. C. Wang. 1995. *Physical Chemistry: Principles and Applications in Biological Sciences*, Prentice Hall, Upper Saddle River, NJ.

Williams, V. R., W. L. Mattice, and H. B. Williams. 1978. *Basic Physical Chemistry for the Life Sciences*, W.H. Freeman, San Francisco.

TABLE S5: Surface Tension of Some Liquids

Liquid	Surface Tension (mN/m)	Temperature (°C)
Water	71.97	25
	58.85	100
Acetone	23.7	20
Benzene	28.9	20
n-Hexane	18.4	20

Source: Adapted from Tinoco et al. (1995).

SUSPENDED SOLIDS IN WATER AND WASTEWATER: Total Suspended Solids

Introduction

Total suspended solids (TSS) in water and wastewater are determined by passing a known volume of the sample through a preweighted filter and weighing the remaining solids after drying the filter at 103°C overnight.

Formula

$$\mathrm{TSS} = \frac{m_f - m_i}{V}$$

where

TSS = total suspended solids (mg/L)

m_f = final filter mass + suspended solids after drying at 103°C

m_i = initial filter mass (mg)

V = sample volume (L)

Reference

Ray, B. T. 1995. *Environmental Engineering*, PWS, Boston.

SUSPENDED SOLIDS IN WATER AND WASTEWATER: Volatile Suspended Solids

Introduction

Volatile suspended solids (VSS) in water and wastewater are determined by burning, at 550°C, the residue obtained from TSS determination.

Formula

$$\mathrm{VSS} = \frac{m_f - m_{f550}}{V}$$

where

VSS = volatile suspended solids (mg/L)
m_f = filter mass after drying at 103°C
m_{f550} = final filter mass after drying at 550°C
V = sample volume (L)

Reference

Ray, B. T. 1995. *Environmental Engineering*, PWS, Boston.

T

TEMPERATURE

See LAGOON TEMPERATURE.

TEMPERATURE COEFFICIENT: Microbial activity

See PHELPS EQUATION

Introduction

The reaction rate of a biological process is temperature dependent.

Formula

The reaction rate K_T at any temperature T is given by the following formula:

$$K_T = K_{20^\circ C}\theta^{T-20}$$

where

K_T = reaction rate at temperature T (1/time)
$K_{20^\circ C}$ = reaction rate at 20°C (1/time)
θ = temperature coefficient

For nitrifiers, Downing (1966) reported the following relationship between temperature and reaction rate:

$$K_N = 0.18 \times 1.128^{T-15}$$

Numerical Values

θ varies with the type of biological process. θ values for biological processes were compiled by Eckenfelder (1970) and are shown in Table T1.

TABLE T1: Temperature Coefficients for Biological Processes

Biological Process	θ
Activated sludge	
(Loading rate = < 0.5 kg BOD/kg MLSS · day	1.0
(Loading rate = > 0.5 kg BOD/kg MLSS · day	1.0–1.04
Trickling filters	1.035
Aerobic lagoon	1.035
Aerobic-facultative lagoon	1.07–1.08
Nitrification	1.07–1.12
$NH_4 \rightarrow NO_2$	1.08
$NO_2 \rightarrow NO_3$	1.06
Benthic oxygen uptake	1.06
Zooplankton respiration	1.05

References

Downing, A. L. 1966. Population dynamics in biological systems. Proc. 3rd Int. Conf. Water Pollut. Res., Munich, 1966, Water Pollution Control Federation, Washington, DC.

Eckenfelder, W. W., Jr., 1970. *Water Quality Engineering for Practicing Engineers*, Barnes & Noble, New York.

Jorgensen, S. E., and I. Johnson. 1989. *Principles of Environmental Science and Technology*, Elsevier, Amsterdam.

THYMIDINE INCORPORATION INTO BACTERIA: Conversion into Bacterial Production

See BACTERIAL PRODUCTION.

TOTAL SUSPENDED SOLIDS

See SUSPENDED SOLIDS IN WATER AND WASTEWATER.

TOXICITY OF CHEMICAL MIXTURES: Additive Index

Introduction

The toxicity of mixtures of two or more chemicals may be additive, antagonistic, or synergistic and is related to the toxicity of the chemicals individually or in mixtures. An additive index was proposed to quantify this relationship (Marking, 1977, 1985; Marking and Mauck, 1975).

Formula

The sum S of the toxic effect of a mixture of two chemicals A and B is given by the following formula:

$$S = \frac{A_m}{A_i} + \frac{B_m}{B_i}$$

where

S = sum of toxic effects
A and B = chemicals
A_i and B_i = toxicity (LC_{50}) of individual chemicals
A_m or B_m = toxicity (LC_{50}) in mixture

- When $S \leqslant 1.0$, additive index $= (1/S) - 1.0$.
- When $S \geqslant 1.0$, additive index $= S(-1) + 1.0$.
- When $S = 1$, additive index $= 0$.

References

Marking, L. L. 1977. Method for assessing additive toxicity of chemical mixtures, pp. 99–108. In: *Aquatic Toxicology and Hazard Evaluation*, F. L. Mayer and J. L. Hamelink, Eds., American Society for Testing and Materials, Philadelphia.

Marking, L. L. 1985. Toxicity of chemical mixtures, pp.164–176. In: *Fundamentals of Aquatic Toxicology*, G. M. Rand and S. R. Petrocelli, Eds., Hemisphere, New York.

Marking, L. L., and W. L. Mauck. 1975. Toxicity of paired mixtures of candidate forest insecticides to rainbow trout. Bull. Environ. Contam. Toxicol. 13: 518–523.

TOXIC UNITS

Definition

The toxic unit acute (TU_a) is the reciprocal of a toxic effluent dilution that causes an acute effect (e.g., 50% mortality) by the end of the acute exposure period. It has been used to sum up the toxicity of individual toxicants in a waste (Lloyd, 1961).

Formula

$$TU_a = \frac{100}{LC_{50}}$$

where LC_{50} is the sample dilution giving 50% mortality (%).

Toxic units have been used to measure the toxic loading rate (LR) as regards the toxic effluent flow into a receiving water (Cary and Barrows, 1981; Mount, 1968; Tebo, 1986):

$$LR = TU_a Q_w = \frac{100 Q_w}{LC_{50}}$$

where

LR = toxic loading rate (gallons/day)

Q_w = daily waste flow rate (gallons/day)

The *toxic unit chronic* (TU_c) is the reciprocal of a toxic effluent dilution that causes no observable effect on the test organism by the end of the chronic exposure period.

$$TU_c = \frac{100}{NOEC}$$

where NOEC = (no observable effect concentration) is the highest dose of the chemical tested that does not produce a significant effect on the test organism (%).

References

Cary, G. A., and M. E. Barrows. 1981. Reduction of toxicity to aquatic organisms by industrial wastewater treatment, EPA-600/53-81-043, U.S. Environmental Protection Agency, Washington, EC.

Lloyd, R. 1961. The toxicity of mixtures of zinc and copper sulphates to rainbow trout (*Salmo gairdneri*, Richardson). Ann. Appl. Biol. 49: 535–538.

Mount, D. I. 1968. Chronic toxicity of copper to fathead minnows (*Pimephales promelas*, Rafinesque). Water Res. 2: 215–233.

Tebo, L. B., Jr. 1986. Effluent monitoring: historical perspective, pp. 13–31. In: *Environmental Hazard Assessment of Effluents*, H. L. Bergman, R. A. Maki, and A. W. Maki, Eds., Pergamon Press, Elmsford, NY.

U.S. EPA. 1985. *Technical Support Document for Water Quality-Based Toxics Control*, EPA-440/4-85-032, Office of Water Regulation and Standards, Washington, DC.

TRICKLING FILTERS

See OXYGEN TRANSFER IN TRICKLING FILTERS; RECIRCULATION RATIO: Trickling Filters; VELZ'S EQUATION.

TRICKLING FILTERS: Substrate Removal

Introduction

A trickling filter is a wastewater treatment process based on attached microbial growth. Microorganisms attach to the filter medium and form a biofilm that is involved in the biotreatment of the incoming wastewater. As regards BOD removal, there is an empirical formula that relates the substrate removal to the filter depth and hydraulic loading (Figure T1).

Formula

$$\frac{S_e}{S_i} = e^{-KZ/(Q/H)^n}$$

where

S_e = substrate concentration in effluent (g/m^3)
S_i = substrate concentration in influent (g/m^3)
K = substrate removal coefficient
Z = filter depth (m)
Q = flow rate (m^3/day)
H = cross-sectional area of filter (m^2)
Q/H = hydraulic loading intensity ($m^3/\text{day} \cdot m^2$)
n = exponent (varies from 0.2 to 1.1)

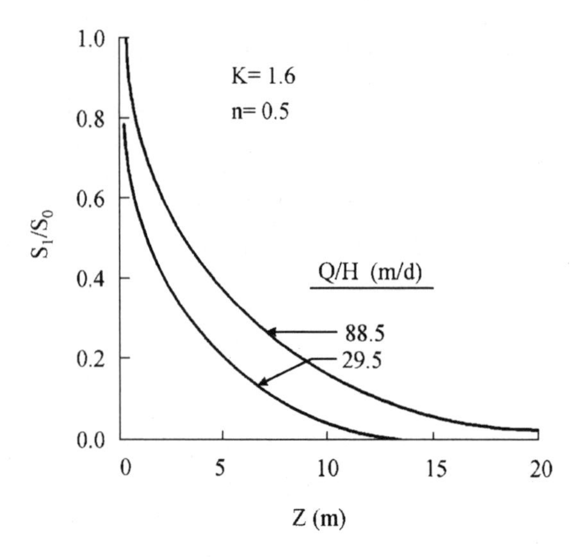

FIGURE T1: Influence of Trickling Filter Depth and Hydraulic Loading Intensity of Substrate Removal rate (*Source:* Data from Rich, 1980; Shulze, 1960.)

K is influenced by several factors, including temperature, substrate type, and concentration. The influence of temperature on K is expressed as

$$K_T = K_{20}\theta^{T-T_{20}} \quad \text{(where } \theta \text{ is 1.035 for organic carbon removal)}$$

References

Rich, L. G. 1980. *Low-Maintenance, Mechanically Simple Wastewater Treatment Systems*, McGraw-Hill, New York.

Schulze, K. L. 1960. Load and efficiency of trickling filters. J. Water Pullut. Control Fed. 32: 245–261.

V

VAND EQUATION

Introduction

The Vand equation describes the rheology of cultures of unicellular microorganisms. It states that the viscosity of a microbial culture is a function of cell density.

Formula

$$\mu = \mu_L(1 + 2.5f_c + 7.25f_c^2)$$

where

μ_L = viscosity of suspending liquid ($kg/m \cdot s$)
f_c = volume fraction of cells

Reference

Trilli, A. 1986. Scale-up fermentations, pp. 277–307. In: *Manual of Industrial Microbiology and Biotechnology*, A. L. Demain and N. A. Solomon, Eds., American Society for Microbiology, Washington, DC.

VELZ'S EQUATION

Introduction

Velz's equation deals with the oxidation of organic matter as a function of filter depth in a trickling filter.

Formula

$$\frac{L_D}{L} = 10^{-KD}$$

where

L_D = remaining removable BOD at depth D (mg/L)
L = total amount of removable BOD in feed (mg/L)

K = reaction constant

D = filter depth (m)

Reference

Hawkes, H. A. 1983. The applied significance of ecological studies of aerobic processes, pp. 173–333. In: *Ecological Aspects of Used-Water Treatment*, vol. 3, *The Processes and Their Ecology*, C. R. Curd and H. A. Hawkes, Eds., Academic Press, London.

VERTICAL EXTINCTION COEFFICIENT

See LIGHT: Vertical Extinction Coefficient.

VOID RATIO, SOIL

See POROSITY, SOIL.

VOLATILE SUSPENDED SOLIDS

See SUSPENDED SOLIDS IN WATER AND WASTEWATER: Volatile Suspended Solids.

W

WATER ACTIVITY

Definition

Water activity (a_w) is the fraction of total water molecules available to microorganisms. Pure water has the maximum water availability ($a_w = 1.00$) and a water potential $\Psi = 0$. Solutions display an a_w value less than 1.00 and a negative Ψ.

Formula

There is a relationship between a_w and the osmolality of the surrounding medium.

$$a_w = \frac{55.51}{vmo + 55.51}$$

where

v = number of ions generated/mole solute
m = molality of solute
o = molal osmotic coefficient
55.51 = number of moles/kg of water

Numerical Values

According to Brown (1990), Jennings (1990), and Kushner (1993),

- Optimum value of a_w: >0.99.
- a_w of pure water is 1.00.
- Saturated solution of NaCl: $a_w = 0.75$.
- Lowest a_w at which growth is possible: $a_w = 0.60$.

For the a_w limits for growth of selected microorganisms, see Table W1.

References

Brown, A. D. 1990. *Microbial Water Stress Physiology: Principles and Perspectives*, Wiley, New York.

TABLE W1: Water Activity a_w Limits for Growth of Selected Microorganisms

	Water Activity, a_w	
Organism	Optimum	Minimum
Bacteria		
Aerobacter aerogenes	<0.99	0.94
Bacillus cereus var. *mycoides*	<0.99	0.99
Bacillus megaterium	<0.99	0.95
Bacillus subtilis	<0.99	0.90
Halococcus sp.	0.99	0.75
Halobacterium sp.	0.90	0.75
Micrococcus sp.	n.d.	0.83
Micrococcus halodenitrificans	0.96	0.83
Pseudomonas fluorescens	0.99	0.97
Staphylococcus aureus	0.99	0.83
Algae		
Dunaliella tertiolecta	<0.99	0.95
Dunaliella viridis	0.94	0.75
Yeasts		
Saccharomyces rouxii	<0.99	0.86
Saccharomyces cerevisiae	n.d.	0.92
Torula utilis	n.d.	0.94
Fungi		
Aspergillus amstelodami	0.94	0.70
Aspergillus flavus	0.98	0.78
Aspergillus nidulans	0.97	0.78
Aspergillus niger	0.98	0.84

Source: Adapted from Reid (1980).

Costilow, R. N. 1981. Biophysical factors in growth, pp. 66–78. In: *Manual of Methods for General Bacteriology*, P. Gerhardt et al., Eds., American Society for Microbiology, Washington, DC.

Jennings, D. H. 1990. Osmophiles, pp. 117–146. In: *Microbiology of Extreme Environments*, Open University Press, Milton Keynes, Buckinghamshire, England.

Kushner, D. J. 1993. Microbial life in extreme environments, pp. 383–407. In: *Aquatic Microbiology: An Ecological Approach*, T. E. Ford, Ed., Blackwell Scientific, Oxford.

Prior, B. A. 1979. Measurement of water activity in foods. J. Food Prot. 42: 668–674.

Reid, D. S. 1980. Water activity as the criterion of water availability, pp. 15–27. In: *Contemporary Microbial Ecology*, D. C. Ellwood, J. N. Hedger, M. J. Latham, J. M. Lynch, and J. H. Slater, Eds. Academic Press, London.

Scott, W. J. 1957. Water relations of food spoilage microorganisms. Adv. Food Res. 3: 84–123.

Smith, D. W. 1982. Extreme natural environments, pp. 555–579. In: *Experimental Microbial Ecology*, R. G. Burns and J. H. Slater, Eds. Blackwell Scientific, Oxford.

WATER BUDGET OF A SYSTEM

Introduction

The sum of water inputs (i.e., precipitation) into a system is equal to the sum of water outputs (i.e., evapotranspiration, runoff, infiltration, storage).

Formula

$$P = ET + R + I + S$$

where

P = precipitation (cm)
ET = evapotranspiration (cm)
R = runoff (cm)
I = infiltration (cm)
S = storage (cm)

Reference

Ray, B. T. 1995. *Environmental Engineering*, PWS, Boston.

WATER CONTENT, SOIL

Formulas

The amount of water in a given soil can be expressed on a volume or on a mass basis:

- *As a volume fraction:* According to Marshall and Holmes (1988), "it is the ratio of the depth of water to the depth of the soil profile that contains it."

$$\theta = \frac{V_l}{V_t}$$

where

V_l = volume of the liquid phase

V_t = total volume in undisturbed sample

- *On a mass basis*

$$\theta_m = \frac{m_w}{m_s}$$

where

m_w = mass of water

m_s = mass of solids

The following formula converts the water content from the mass basis to the volume fraction:

$$\theta = \frac{\theta_m \rho_b}{\rho_w}$$

where

ρ_b = soil bulk density (g/cm^3)

ρ_w = water density (g/cm^3)

Reference

Marshall, T. J., and J. W. Holmes. 1988. *Soil Physics*, 2nd ed., Cambridge University Press, Cambridge.

WATER EFFECT RATIO (WER)

Introduction

The concept of water effect ratio (WER) was proposed by the U.S. EPA (1982, 1984), and an interim guidance was published (U.S. EPA, 1994). This ratio helps adjust the LC_{50} for toxic metals as a function of the site-specific physical and chemical charactertistics, such as pH, hardness, salinity, presence of chelating agents (e.g., humic substances), and suspended particulates in receiving waters. WER needs to be determined for each metal at each site.

Formula

WER is the ratio of the LC_{50} derived from testing the toxicity of a metal to fish or invertebrates using the site water to the LC_{50} derived from testing the toxicity of the same metal to the same indicator organism in standard laboratory water:

$$\mathrm{WER} = \frac{\text{site water } LC_{50}}{\text{lab water } LC_{50}}$$

This ratio takes into account biological, physical, and chemical characteristics of the site water. If the water effect ratio for a particular metal is significantly different from 1.0, the national water quality criterion for that metal should be multiplied by the WER to obtain the site-specific criterion for the given metal. For a given site, WER cannot be extrapolated from one metal to another.

References

U.S. EPA. 1982. *Water Quality Standards Handbook* (draft), Office of Water Regulations and Standards, Washington, DC.

U.S. EPA. 1984. *Guidelines for Deriving Numerical Aquatic Site-Specific Water Quality Criteria by Modifying National Criteria*. EPA-600/384009, Environmental Research Laboratory, Duluth, MN.

U.S. EPA. 1994. *Interim Guidance on Determination and Use of Water-Effect Ratios for Metals*, EPA-823-B-94-01, Office of Science and Technology, Washington, DC.

WATER POTENTIAL, SOIL

Introduction

Availability of water to microorganisms in soils is expressed as water activity (*See* WATER ACTIVITY) or as soil water potential. Water potential describes the force that holds water in soils. Microorganisms require energy to use the water held in the soil.

Formula

$$\Psi = \frac{RT}{V_w} \ln a_w$$

where

Ψ = water potential (Pa)

R = universal gas constant

T = absolute temperature (K)

V_w = partial molal volume of water ($18\,\text{cm}^3/\text{mol}$ at 4°C)

a_w = water activity, or the relative humidity of the soil atmosphere ($a_w = P/P_0$; P is the vapor pressure of soil air and P_0 is the vapor pressure of saturated air at the same temperature as the soil air)

References

Metting, F. B. Jr. 1993. Structure and physiological ecology of soil microbial communities. In: *Soil Microbial Ecology*, F. B. Metting, Jr., Ed., Marcel Dekker, New York.

Papendick, R. I., and G. S. Campbell. 1981. Theory and measurement of water potential, pp. 1–22. In: *Water Potential Relations in Soil Microbiology*, J. F. Parr, W. R. Gardner, and L. F. Elliot, Eds., Soil Science Society of America, Madison, WI.

Skujins, J. 1984. Microbial ecology of desert soils. Adv. Microb. Ecol. 7: 49–91.

WATSON LAW

See DISINFECTION: Lethality Coefficient.

WOOLF–AUGUSTINSSON–HOFSTEE PLOT

See HOFSTEE EQUATION.

Y

YIELD

See GROWTH YIELD OF MICROORGANISMS.

YOUNG'S EQUATION

Introduction

Young's equation gives the contact angle of a liquid with a solid surface. It determines the wettability of a solid surface.

Formula

$$\gamma_{SV} = \gamma_{SL} + \gamma_{LV} \cos \theta$$

where

γ_{SV} = interfacial tension at solid–vapor interface (dyne/cm)
γ_{SL} = interfacial tension at solid–liquid interface (dyne/cm)
γ_{LV} = interfacial tension at liquid–vapor interface (dyne/cm)
θ = contact angle

References

Fletcher, M., and K. C. Marshall. 1982. Are solid surfaces of ecological significance to aquatic bacteria? Adv. Mirob. Ecol. 6: 199–235.

Marshall, K. C. 1976. *Interfaces in Microbial Ecology*, Harvard University Press, Cambridge, MA.

Z

ZETA POTENTIAL

See also ELECTRICAL DOUBLE LAYER: Thickness; SMOLUCHOWSKI EQUATION.

Introduction

Zeta potential is the potential at the surface of a colloidal particle. It is important in predicting the coagulation of colloidal particles in water and wastewater treatment. It is necessary to neutralize the charge of the colloidal particle to induce their flocculation and precipitation. Addition of electrolytes (e.g., addition of aluminum sulfate) or polyelectrolytes results in a decrease of the zeta potential, leading to a decrease in repulsion forces between particles.

Formulas

The zeta potential ζ is given by (Eckenfelder, 1989)

$$\zeta = \frac{4\pi\eta \times \text{EM}}{\varepsilon}$$

where

ζ = zeta potential (mV)
π = 3.1416
η = viscosity of the medium (poise)
EM = electrophoretic mobility (μm/s · V · cm) = v/X [where v = particle velocity (μm/s) and X = applied potential per unit length of cell (V/cm)]
ε = dielectric constant of the liquid

A practical formula for ζ is

$$\zeta(\text{mV}) = \frac{113{,}000}{\varepsilon}\eta \times \text{EM}$$

Zeta potential can be estimated by measuring the movement of colloidal particles in an electric field and determining the electrophoretic mobility of the particles. The zeta potential at 25°C can be obtained via

$$\zeta = 12.8\mathrm{EM}$$

Numerical Values

The average ζ for water and wastewater colloidal particles is -16 to -22 mV. The zeta potential reported for 19 bacteria strains suspended in deionized water varied from -8 to -36 mV (Gannon et al., 1991).

References

Eckenfelder, W. W., Jr. 1989. *Industrial Water Pollution Control*, 2nd ed., McGraw-Hill, New York.

Gannon, J. T., V. B. Manilal, and M. Alexander. 1991. Relationship between cell surface properties and transport of bacteria through soil. Appl. Environ. Microbiol. 57: 190–193.

Marshall, K. C. 1976. *Interfaces in Microbial Ecology*, Harvard University Press, Cambridge, MA.

Sawyer, C. N., P. L. McCarty, and G. F. Parkin. 1994. *Chemistry for Environmental Engineering*, McGraw-Hill, New York.

ZOOPLANKTON

See CLEARANCE RATE: Zooplankton and Protozon; EGESTION (EVACUATION) RATE OF ZOOPLANKTON; FILTERING RATE OF ZOOPLANKTON; INGESTION RATE OF ZOOPLANKTON.

APPENDICES

APPENDICES

Appendix 1: Système International (SI Units)

Table 1A: Base Units in the SI System
Table 1B: Units Derived from the SI System
Table 1C: Prefixes Used in the SI System

Appendix 2: Greek Alphabet

Appendix 3: Some Useful Conversions

Appendix 4: Some Useful Constants

Appendix 5: Temperature Conversions

Appendix 6: Radiobiology

Table 6A: Units Used in Radiobiology
Table 6B: Half-Life and Type of Particles Emitted by Selected Radioisotopes

Appendix 7: Light

Table 7A: Approximate Wavelength Ranges for the Various Regions of the Electromagnetic Spectrum
Table 7B: Conversions for Light

Appendix 8: Physical Properties of Water

Appendix 9: Solubility of Oxygen in Water

Appendix 10. Conversion of ppm, ppb, and ppt of Chemicals to Concentrations Expressed in SI Units

Appendix 11. International Atomic Weights

APPENDIX 1
Système International (SI Units)

Système International (SI) is an international system that has been recommended by the International Organization for Standardization. It is a system of units designed to standardize scientific measurements. It has been adopted by many countries around the world.

TABLE 1A: Base Units in the SI System

Quantity	Base SI Unit	Symbol
Length	meter	m
Mass	kilogram	kg
Time	second	s
Electric current	ampere	A
Temperature	kelvin	K
Amount of substance	mole	mol
Luminous intensity	candela	cd

TABLE 1B: Units Derived from the SI System

Measurement	Derived SI Unit	Symbol	Equivalent SI Unit
Volume	liter	L	$1\ L = 10^{-3}\ m^3$
Force	newton	N	$1\ N = kg \cdot m \cdot s^2$
Energy, work	joule	J	$1\ J = N \cdot m$
Power	watt	W	$1\ W = 1\ J/s$
Frequency	hertz	Hz	$1\ Hz = s^{-1}$
Pressure	pascal	Pa	$1\ Pa = 1\ N \cdot m^{-2} = 1\ kg/m \cdot s^2$
Viscosity	centipoise	cp	$1\ cp = 10^{-3}\ kg/m \cdot s$
Electric charge, quantity of electricity	coulomb	C	$1\ C = 1\ A \cdot s$
Electric potential	volt	V	$1\ V = 1\ J/C = 1\ W/A$
Electric capacitance	farad	F	$1\ F = 1\ C/V$
Electric resistance	ohm	Ω	$1\ \Omega = 1\ V/A$
Electric conductance	siemens	S	$1\ S = 1\ \Omega^{-1} = 1\ A/V$
Magnetic flux	weber	Wb	$1\ Wb = 1\ V \cdot s$
Magnetic flux density	tesla	T	$1\ T = 1\ V \cdot s/m^2$
Activity (of ionizing radiation source)	becquerel	Bq	s^{-1}

TABLE 1C: Prefixes Used in the SI System

Prefix	Symbol	Multiples and submultiples
yotta	Y	10^{24}
zetta	Z	10^{21}
exa	E	10^{18}
peta	P	10^{15}
tera	T	10^{12}
giga	G	10^{9}
mega	M	10^{6}
kilo	k	10^{3}
hecto	h	10^{2}
deka	da	10^{1}
deci	d	10^{-1}
centi	c	10^{-2}
milli	m	10^{-3}
micro	μ	10^{-6}
nano	n	10^{-9}
pico	p	10^{-12}
femto	f	10^{-15}
atto	a	10^{-18}
zepto	z	10^{-21}
yocto	y	10^{-24}

APPENDIX 2
Greek Alphabet

A	α	alpha
B	β	beta
Γ	γ	gamma
Δ	δ	delta
E	ε	epsilon
Z	ζ	zeta
H	η	eta
Θ	θ	theta
I	ι	iota
K	κ	kappa
Λ	λ	lambda
M	μ	(mu)
N	ν	(nu)
Ξ	ξ	xi
O	o	omikron
Π	π	pi
P	ρ	rho
Σ	σ	sigma
T	τ	tau
Υ	υ	upsilon
Φ	ϕ	phi
X	χ	chi
Ψ	ψ	psi
Ω	ω	omega

APPENDIX 3: Some Useful Conversions

To Convert (Symbol):	Multiply by:	To Obtain (Symbol):
Length		
inch (in)	2.54	centimeter (cm)
inch	0.0254	meter (m)
foot (ft)	30	centimeter
yard (yd)	0.9144	meter
meter	1.094	yard
mile (mi)	1.61	kilometer (km)
mile	5280	foot
centimeter	0.39	inch
meter	39.37	inch
meter	1.094	yard
meter	3.3	foot
kilometer	0.62	mile
Mass		
ounce (oz)	28.3	gram (g)
pound (lb)	0.453	kilogram (kg)
pound	453.6	gram
kilogram (kg)	2.2	pound (lb)
ton	1000	kilogram
ton (short)	907.2	kilogram
Volume		
quart (qt)	0.946	liter (L)
pint (pt)	0.47	liter
fluid ounce (fl oz)	0.03	liter
gallon (gal)	3.785	liter
cubic foot (ft^3)	28.316	liter
cubic inch	0.01639	liter
liter	1000	milliliter (mL)
liter	2.1	pint
liter	0.26	gallon
liter	30	fluid ounce
cubic meter (m^3)	1000	liter
cubic meter	1.31	cubic yard (yd^3)
cubic meter	35.31	cubic foot (ft^3)

Appendix 3 (*Continued*)

To Convert (Symbol):	Multiply by:	To Obtain (Symbol):
Surface		
square inch (in^2)	6.452	square centimeter (cm^2)
square foot (ft^2)	0.0929	square meter (m^2)
square foot	929.0	square centimeter
acre	0.4047	hectare (ha)
acre	4,047	square meter
square mile (mi^2)	2.590	square kilometer (km^2)
square centimeter (cm^2)	1×10^{-4}	square meter
square centimeter	1.076×10^{-3}	square feet
square meter	1×10^{-4}	hectare
square meter	1×10^{4}	square centimeter
square meter	10.76	square feet
hectare	2.471	acre
hectare	1.076×10^{5}	square foot
hectare	1×10^{4}	square meter
Pressure		
Pound/square inch (lb/in^2)	6.8948×10^{3}	pascal (Pa) or newton/m^2
pound/square foot (lb/ft^2)	47.8803	pascal
Newton/m^2	1	pascal
atmosphere	1.033	kilograms/cm^2
atmosphere	1.013	bar
bar	1×10^{6}	dynes/cm^2
Flow rate		
gallon/day (gal/d)	4.3813×10^{-2}	liter/second (L/s)
gallon/minute (gal/min)	6.3090×10^{-2}	liter/second
million gallons/day	43.8126	liter/second
(Mgal/d)	3.7854×10^{3}	m^3/day
million gallons/day	5.886×10^{-4}	cubic feet/second
liter/minute (L/min)		
Power		
horsepower (hp)	0.7457	kilowatt (kW)
kilowatt	1.341	horsepower
Acceleration		
foot/second^2 (ft/s^2)	0.3048	meter/second^2 (m/s^2)
inch/second^2	0.0254	meter/second^2
Charge		
coulomb	3×10^{9}	electrostatic units (esu)

Appendix 3 (*Continued*)

To Convert (Symbol):	Multiply by:	To Obtain (Symbol):
Energy		
kilowatt-hour (kWh)	3600	kilojoule (kJ)
watt-hour	3.600	joule (J)
joule	1×10^7	erg or $g \cdot cm^2/s^2$
joule	2.389×10^{-4}	kilocalorie (kcal)
calorie	4.184	joules
Force		
newton	1	kilogram $\cdot$ m/s^2
newton	10^5	dynes
dyne	1	$g \cdot cm/s^2$
Temperature		
Fahrenheit (°F)	0.555 (°F − 32)	Celsius (°C)
Celsius (°C)	(1.8°C) + 32	Fahrenheit (°F)

Source: Data from O. T. Lind, *Handbook of Common Methods in Limnology*, 2nd ed., C. V. Mosby, St. Louis, MO, 1979; M. B. McBride, *Environmental Chemistry of Soils*, Oxford University Press, New York, 1994; L. G. Rich, *Low-Maintenance, Mechanically Simple Wastewater Treatment Systems*, McGraw-Hill, New York, 1980.

APPENDIX 4: Some Useful Constants

Atmospheric pressure	atm = 1,013,250 dyn/cm^2
Atom mass	$m_u = 1.6605402 \times 10^{-27}$ kg
Avogadro's number	$N = 6.0221367 \times 10^{23}$ mol^{-1}
Boltzmann constant $k = R/N$	$k = 1.380658 \times 10^{-23}$ J/K
Elementary charge	$e = 1.60217733 \times 10^{-19}$ C
Faraday's constant	$F = 9.6485309 \times 10^4$ C/mol
Gas constant	$R = kN = 8.314510$ J/mol · K
Gravitational acceleration	$g = 9.80665$ m/s^2
Mass of an electron at rest	$m_e = 9.109 \times 10^{-31}$ kg
Molar volume of an ideal gas at 1 atm and 25°C	$\acute{U}_{ideal\ gas} = 24.465$ L/mol
Permittivity of vacuum	$e_0 = 8.854187 \times 10^{-12}$ C/V · m
Planck constant	$h = 6.6260755 \times 10^{-34}$ J · s
Proton mass	$m_p = 1.6726 \times 10^{-27}$ kg
Speed of light in *vacuo*	$c = 2.99792 \times 10^8$ m/s
Stefan–Boltzmann constant	$\sigma = 5.67 \times 10^{-8}$ $J/m^2 \cdot s \cdot K^4$
Zero of the Celsius scale	0°C = 273.15 K

Source: Data from C. N. Sawyer and P. L. McCarty, *Chemistry for Environmental Engineering*, McGraw-Hill, New York, 1978; R. P. Schwarzenbach, P. M. Gschwend and D. M. Imboden, *Environmental Organic Chemistry*, Wiley, New York, 1993; V. R. Williams, W. L. Mattice, and H. B. Williams, *Basic Physical Chemistry for the Life Sciences*, 3rd ed., W.H. Freeman, San Francisco, 1978.

APPENDIX 5:
Temperature Conversions

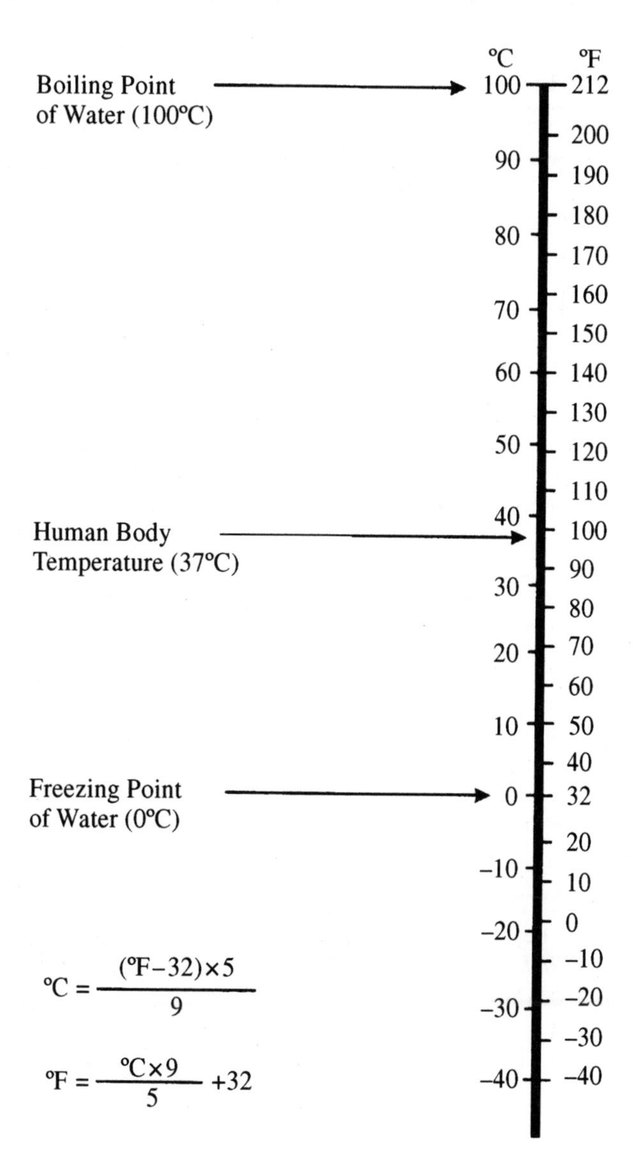

$$°C = \frac{(°F - 32) \times 5}{9}$$

$$°F = \frac{°C \times 9}{5} + 32$$

APPENDIX 6: Radiobiology

TABLE 6A: Units Used in Radiobiology

Unit or Quantity	Symbol	Application
Becquerel	Bq	SI quantity of radioactivity Bq = 1 disintegration/s $Bq = 2.7 \times 10^{-11}$ Ci
Curie	Ci	Quantity of radioactivity $1\ Ci = 3.7 \times 10^{10}$ disintegrations/s $1\ Ci = 3.7 \times 10^{10}$ Bq
Electron volt	eV	Unit of energy $1\ eV = 1.6 \times 10^{-12}$ erg $1\ eV = 1.6 \times 10^{-19}$ J
Linear energy transfer	LET	Energy deposition per unit of path length; usually in eV/mm
Quality factor	Q	Biological effectiveness of radiations
Gray	Gy	SI unit of absorbed dose 1 Gy = 100 rad = 1 J/kg
Rad	rad	Unit of absorbed dose 1 rad = 0.01 Gy = 100 erg/g
Rem	rem	Unit of dose equivalent rad × Q × other modifying factor 1 rem = 0.01 Sv
Vievert	Sv	SI unit of dose equivalent rad × Q × other modifying factor 1 Sv = 100 rem

Source: BEIR V, 1990. Reprinted with permission from *Health Effects of Exposure to Low Levels of Ionizing Radiation (BEIR V)*. Copyright 1990 by the National Academy of Sciences. Courtesy of the National Academy Press, Washington, DC.

TABLE 6B: Half-Life and Type of Particle Emitted by Selected Radioisotopes

Isotope	Half-Life	Type of Particle Emitted
^{14}C	5730 years	β
^{226}Ra	1630 years	α, γ
^{222}Rn	3.82 days	α
^{3}H	12.46 years	β
^{131}I	8.05 days	β, γ
^{32}P	14.3 days	β
^{35}S	89 days	β
^{59}Fe	44.3 days	β, α
^{203}Hg	45.4 days	β, α
^{63}Ni	125 days	β
^{45}Ca	165 days	β, α

Source: Data from *Oak Ridge National Laboratory Catalog of Radio and Stable Isotopes*, Isotope Development Center, Oak Ridge National Laboratory, Oak Ridge, TN, 4th revision, 1963; B. J. C. Efiok, *Basic Calculations for Chemical and Biological Analysis*, AOAC International, Arlington, VA, 1993.

APPENDIX 7: Light

TABLE 7A: Approximate Wavelength Ranges for the Various Regions of the Electromagnetic Spectrum

Name of Radiation	Wavelength Range
γ rays	0.003–0.3 Å
X-rays	0.3–100 Å
Far ultraviolet	100–2000 Å
Ultraviolet	200–400 nm
Visible	400–800 nm
Near infrared	0.8–2.5 μm
Infrared	2.5–15 μm
Far infrared	15–200 μm
Microwave	0.2–7.0 mm
Radar	7–100 mm
Very high frequency	10–1000 cm
Ultrahigh frequency	10–100 m
Radio waves	10–10,000 m

Source: V. R. Williams, W. L. Mattice, and H. B. Williams, *Basic Physical Chemistry for the Life Sciences*, 3rd ed. W.H. Freeman, San Francisco, 1978.

TABLE 7B: Conversions for Light

To Convert:	Multiply by:	To Obtain:
einsteins	6.024×10^{23}	quanta
footcandles	1.0	lumens/ft^2
footcandles	10.764	lumens/m^2
footcandles	10.764	lux
lumens	0.318	candellas
lumens/ft^2	1.0	footcandles
lux	0.0929	footcandles
quanta	1.66×10^{-24}	einsteins
candella	3.1416	lumens

APPENDIX 8: Physical Properties of Water

Density (25°C), kg/m^3	997.075
Density (20°C), kg/m^3	998.2
Maximum density, kg/m^3	1000.000
Temperature of maximum density, °C	3.940
Viscosity (25°C), Pa/s	0.890×10^{-3}
Kinematic viscosity (25°C), m^2/s	0.89×10^{-6}
Melting point (101,325 Pa), °C	0.0000
Boiling point (101,325 Pa), °C	100.00
Latent heat of ice, kJ/mol	6.0104
Latent heat of evaporation, kJ/mol	40.66
Specific heat capacity (15°C), J/kg · °C	4186
Thermal conductivity (25°C)m, J/cm · s · °C	0.00569
Heat of vaporization, J/kg	2.453×10^6
Surface tension (25°C), N/m	71.97×10^{-3}
Surface tension (20°C), N/m	72.75×10^{-3}
Surface tension (0°C), N/m	75.64×10^{-3}
Dielectric constant (25°C)	78.54
Vapor pressure (20°C), torr	17.535

Source: Data from J. Schwoerbel, *Handbook of Limnology*, Ellis Horwood, Chichester, West Sussex, England, 1987; T. J. Marshall and J. W. Holmes, *Soil Physics*, 2nd ed., Cambridge Unviversity Press, Cambridge, 1988.

APPENDIX 9:
Solubility of Oxygen in Water

Stated in the table is the solubility of dissolved oxygen in water (mg/L) in equilibrium with dry air at 760 mm Hg and containing 20.9% oxygen

Temperature (°C)	Chloride Concentration (mg/L)				
	0	5000	10,000	15,000	20,000
0	14.6	13.8	13.0	12.1	11.3
1	14.2	13.4	12.6	11.8	11.0
2	13.8	13.1	12.3	11.5	10.8
3	13.5	12.7	12.0	11.2	10.5
4	13.1	12.4	11.7	11.0	10.3
5	12.8	12.1	11.4	10.7	10.0
6	12.5	11.8	11.1	10.5	9.8
7	12.2	11.5	10.9	10.2	9.6
8	11.9	11.2	10.6	10.0	9.4
9	11.6	11.0	10.4	9.8	9.2
10	11.3	10.7	10.1	9.6	9.0
11	11.1	10.5	9.9	9.4	8.8
12	10.8	10.3	9.7	9.2	8.6
13	10.6	10.1	9.5	9.0	8.5
14	10.4	9.9	9.3	8.8	8.3
15	10.2	9.7	9.1	8.6	8.1
16	10.0	9.5	9.0	8.5	8.0
17	9.7	9.3	8.8	8.3	7.8
18	9.5	9.1	8.6	8.2	7.7
19	9.4	8.9	8.5	8.0	7.6
20	9.2	8.7	8.3	7.9	7.4
21	9.0	8.6	8.1	7.7	7.3
22	8.8	8.4	8.0	7.6	7.1
23	8.7	8.3	7.9	7.4	7.0
24	8.5	8.1	7.7	7.3	6.9
25	8.4	8.0	7.6	7.2	6.7
26	8.2	7.8	7.4	7.0	6.6
27	8.1	7.7	7.3	6.9	6.5
28	7.9	7.5	7.1	6.8	6.4
29	7.8	7.4	7.0	6.6	6.3
30	7.6	7.3	6.9	6.5	6.1

Source: G. C. Whipple and M. C. Whipple, Solubility of oxygen in sea water, J. Am. Chem. Soc. 33: 362 (1911).

APPENDIX 10:
Conversion of ppm, ppb, and ppt of Chemicals to Concentrations Expressed in SI Units

Medium	Conversion to SI Units[a]
Water (4°C, 1 atm)	
ppm	1 ppm = 1 g/m^3 = 1 mg/L
ppb	1 ppb = 1 mg/m^3
ppt	1 ppt = 1 $\mu g/m^3$
Air	
ppm	1 ppm = $1 \times \frac{M}{22.4}$ mg/m^3
ppb	1 ppb = $1 \times \frac{M}{22.4}$ $\mu g/m^3$
ppt	1 ppt = $1 \times \frac{M}{22.4}$ ng/m^3
Soil	
ppm	1 ppm = 1 mg/kg soil
ppb	1 ppb = 1 μg/kg soil
ppt	1 ppt = 1 ng/kg soil

Source: Adapted from L. J. Thibodeaux, *Chemodynamics: Environmental Movement of Chemicals in Air, Water, and Soil*, Wiley, New York, 1979.

[a] *M*, molecular weight of chemical (g/mol).

APPENDIX 11: International Atomic Weights

Name	Symbol	Atomic Number	Atomic Weight	Name	Symbol	Atomic Number	Atomic Weight
Actinium	Ac	89	227.0278	Copper	Cu	29	63.546
Aluminium	Al	13	26.981539	Curium	Cm	96	243.0614
Americium	Am	95	241.0568	Dyprosium	Dy	66	162.50
Antimony	Sb	51	121.75	Einsteinium	Es	99	252.083
Argon	Ar	18	39.948	Erbium	Er	68	167.26
Arsenic	As	33	74.92159	Europium	Eu	63	151.965
Astatine	At	85	209.9871	Fermium	Fm	100	257.0951
Barium	Ba	56	137.327	Fluorine	F	9	18.9984032
Berkelium	Bk	97	247.0703	Francium	Fr	87	223.0197
Beryllium	Be	4	9.012182	Gadolinium	Gd	64	157.25
Bismuth	Bi	83	208.98037	Gallium	Ga	31	69.723
Boron	B	5	10.811	Germanium	Ge	32	72.61
Bromine	Br	35	79.904	Gold	Au	79	196.96654
Cadmium	Cd	48	112.411	Hafnium	Hf	72	178.49
Calcium	Ca	20	40.078	Helium	He	2	4.0026002
Californium	Cf	98	249.0748	Holmium	Ho	67	164.93032
Carbon	C	6	12.011	Hydrogen	H	1	1.00794
Cerium	Ce	58	140.115	Indium	In	49	114.82
Cesium	Cs	55	132.90543	Iodine	I	53	126.9044
Chlorine	Cl	17	35.4527	Iridium	Ir	77	192.22
Chromium	Cr	24	51.9961	Iron	Fe	26	55.847
Cobalt	Co	27	58.93320	Krypton	Kr	36	83.80

Lanthanum	La	57	138.9055	Potassium	K	19	39.0983
Lawrencium	Lr	103	262.11	Praseodymium	Pr	59	140.90765
Lead	Pb	82	207.2	Promethium	Pm	61	144.9127
Lithium	Li	3	6.941	Protactinium	Pa	91	231.0359
Lutetium	Lu	71	174.967	Radium	Ra	88	223.0185
Magnesium	Mg	12	24.3050	Radon	Rn	86	210.8806
Manganese	Mn	25	54.93805	Rhenium	Re	75	186.207
Mendelevium	Md	101	256.094	Rhodium	Rh	45	102.90550
Mercury	Hg	80	200.59	Rubidium	Rb	37	85.4678
Molybdenum	Mo	42	95.94	Ruthenium	Ru	44	101.07
Neodymium	Nd	60	144.24	Samarium	Sm	62	150.36
Neon	Ne	10	20.1797	Scandium	Sc	21	44.955910
Neptunium	Np	93	237.0482	Selenium	Se	34	78.96
Nickel	Ni	28	58.69	Silicon	Si	14	28.0855
Niobium	Nb	41	92.90638	Silver	Ag	47	107.8682
Nitrogen	N	7	14.00674	Sodium	Na	11	22.989768
Nobelium	No	102	259.1009	Stontium	Sr	38	87.62
Osmium	Os	76	190.2	Sulfur	S	6	32.066
Oxygen	O	8	15.9994	Tantalum	Ta	73	180.9479
Palladium	Pd	46	106.42	Technetium	Tc	43	96.9064
Phosphorus	P	15	30.973762	Tellurium	Te	52	127.60
Platinum	Pt	78	195.08	Terbium	Tb	65	158.92534
Plutonium	Pu	94	238.0496	Thallium	Tl	81	204.3833
Polonium	Po	84	208.9824	Thorium	Th	90	232.0381

(continued)

Appendix 11 (*Continued*)

Name	Symbol	Atomic Number	Atomic Weight	Name	Symbol	Atomic Number	Atomic Weight
Thulium	Tm	69	168.93421	Uranium	U	92	238.0289
Tin	Sn	50	118.710	Vanadium	V	23	50.9415
Titanium	Ti	22	47.88	Xenon	Xe	54	131.29
Tungsten	W	74	183.85	Ytterbium	Yb	70	173.04
Unnilhexium	Unh	106	263.118	Yttrium	Y	39	88.90585
Unnilpentium	Unp	105	262.114	Zinc	Zn	30	65.39
Unnilquadium	Unq	104	261.11	Zirconium	Zr	40	91.224
Unnilseptium	Uns	107	262.12				

ENVIRONMENTAL SCIENCE AND TECHNOLOGY

A Wiley-Interscience Series of Texts and Monographs

Edited by JERALD L. SCHNOOR, *University of Iowa*
ALEXANDER ZEHNDER, *Swiss Federal Institute for Water Resources and Water Pollution Control*

PHYSIOCHEMICAL PROCESSES FOR WATER QUALITY CONTROL
Walter J. Weber, Jr., Editor

pH AND pION CONTROL IN PROCESS AND WASTE STREAMS
F. G. Shinskey

AQUATIC POLLUTION: An Introductory Text
Edward A. Laws

INDOOR AIR POLLUTION: Characterization, Prediction, and Control
Richard A. Wadden and Peter A. Scheff

PRINCIPLES OF ANIMAL EXTRAPOLATION
Edward J. Calabrese

SYSTEMS ECOLOGY: An Introduction
Howard T. Odum

INTEGRATED MANAGEMENT OF INSECT PESTS OF POME AND STONE FRUITS
B. A. Croft and S. C. Hoyt, Editors

WATER RESOURCES: Distribution, Use and Management
John R. Mather

ECOGENETICS: Genetic Variation in Susceptibility to Environmental Agents
Edward J. Calabrese

GROUNDWATER POLLUTION MICROBIOLOGY
Gabriel Bitton and Charles P. Gerba, Editors

CHEMISTRY AND ECOTOXICOLOGY OF POLLUTION
Des W. Connell and Gregory J. Miller

SALINITY TOLERANCE IN PLANTS: Strategies for Crop Improvement
Richard C. Staples and Gary H. Toenniessen, Editors

ECOLOGY, IMPACT ASSESSMENT, AND ENVIRONMENTAL PLANNING
Walter E. Westman

CHEMICAL PROCESSES IN LAKES
Werner Stumm, Editor

INTEGRATED PEST MANAGEMENT IN PINE-BARK BEETLE ECOSYSTEMS
William E. Waters, Ronald W. Stark, and David L. Wood, Editors

PALEOCLIMATE ANALYSIS AND MODELING
Alan D. Hecht, Editor

BLACK CARBON IN THE ENVIRONMENT: Properties and Distribution
E. D. Goldberg

GROUND WATER QUALITY
C. H. Ward, W. Giger, and P. L. McCarty, Editors

TOXIC SUSCEPTIBILITY: Male/Female Differences
Edward J. Calabrese

ENERGY AND RESOURCE QUALITY: The Ecology of the Economic Process
Charles A. S. Hall, Cutler J. Cleveland, and Robert Kaufmann

AGE AND SUSCEPTIBILITY TO TOXIC SUBSTANCES
Edward J. Calabrese

ENVIRONMENTAL SCIENCE AND TECHNOLOGY

List of Titles (*Continued)*

ECOLOGICAL THEORY AND INTEGRATED PEST MANAGEMENT PRACTICE
Marcos Kogan, Editor

AQUATIC SURFACE CHEMISTRY: Chemical Processes at the Particle Water Interface
Werner Stumm, Editor

RADON AND ITS DECAY PRODUCTS IN INDOOR AIR
William W. Nazaroff and Anthony V. Nero, Jr., Editors

PLANT STRESS–INSECT INTERACTIONS
E. A. Heinrichs, Editor

INTEGRATED PEST MANAGEMENT SYSTEMS AND COTTON PRODUCTION
Ray Frisbie, Kamal El-Zik, and L. Ted Wilson, Editors

ECOLOGICAL ENGINEERING: An Introduction to Ecotechnology
William J. Mitsch and Sven Erik Jorgensen, Editors

ARTHROPOD BIOLOGICAL CONTROL AGENTS AND PESTICIDES
Brian A. Croft

AQUATIC CHEMICAL KINETICS: Reaction Rates of Processes in Natural Waters
Werner Stumm, Editor

GENERAL ENERGETICS: Energy in the Biosphere and Civilization
Vaclav Smil

FATE OF PESTICIDES AND CHEMICALS IN THE ENVIRONMENT
J. L. Schnoor, Editor

ENVIRONMENTAL ENGINEERING AND SANITATION, Fourth Edition
Joseph A. Salvato

TOXIC SUBSTANCES IN THE ENVIRONMENT
B. Magnus Francis

CLIMATE-BIOSPHERE INTERACTIONS
Richard G. Zepp, Editor

AQUATIC CHEMISTRY: Chemical Equilibria and Rates in Natural Waters, Third Edition
Werner Stumm and James J. Morgan

PROCESS DYNAMICS IN ENVIRONMENTAL SYSTEMS
Walter J. Weber, Jr., and Francis A. DiGiano

ENVIRONMENTAL CHEMODYNAMICS: Movement of Chemicals in Air, Water, and Soil, Second Edition
Louis J. Thibodeaux

ENVIRONMENTAL MODELING: Fate and Transport of Pollutants in Water, Air, and Soil
Jerald L. Schnoor

TRANSPORT MODELING FOR ENVIRONMENTAL ENGINEERS AND SCIENTISTS
Mark M. Clark

FORMULA HANDBOOK FOR ENVIRONMENTAL ENGINEERS AND SCIENTISTS
Gabriel Bitton

6273

REFERENCE BOOK
NOT TO BE TAKEN
FROM THE LIBRARY